D0204026

Cake-Cutting Algorithms

Cake-Cutting Algorithms

Be Fair If You Can

Jack Robertson, 1937-
William Webb, 1944-

A K Peters
Natick, Massachusetts

Editorial, Sales, and Customer Service Office

A K Peters, Ltd.
63 South Avenue
Natick, MA 01760

Library of Congress Cataloging-in-Publication Data

Robertson, Jack, 1937–
Cake-cutting algorithms: be fair if you can /
Jack Robertson, William Webb.
p. cm.
Includes bibliographical references and index.
ISBN 1-56881-076-8
1. Cutting stock problem. 2. Fairness. 3. Algorithms. I. Webb, William, 1944– . II. Title.
T57.6.R63 1998
511′.64—dc21 97-41258
CIP

Printed in the United States of America
02 01 00 99 98 10 9 8 7 6 5 4 3 2 1

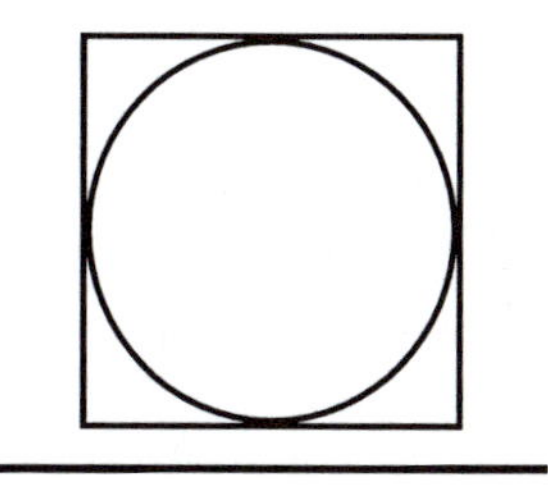

Contents

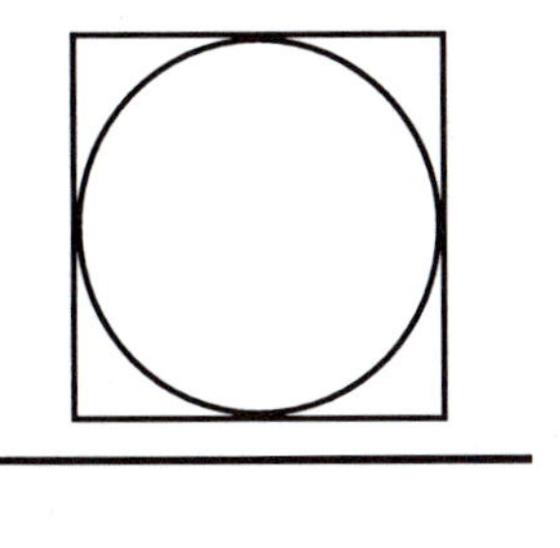

Preface

Why a Book on Cake-Cutting Algorithms?

Since the famous Polish school of mathematicians (Steinhaus, Banach, and Knaster) introduced the problem of fair division and presented the first algorithms in the 1940's, the problem has blossomed and been widely popularized. There are a large number of results scattered in research journals. Different interpretations of "fair" have been proposed, and various classes of algorithms employed. More recently, some impossibility and optimality results address the question raised by Steinhaus at the outset: How many cuts are required for fair division? Although serious mathematical attention to the problem is fairly recent, it is becoming widely exposed as results in the area are presented in popular periodicals, elementary courses, and textbooks. In the numerous opportunities we have had to discuss this material in a variety of settings, it has been our experience that people quickly take up the challenge of searching for fair division methods. It seems to us that the time has come to have the algorithms for fair division collected in one readable, inclusive source.

We have attempted to give as comprehensive an answer as we could to the question: What is known about cake-cutting algorithms using various definitions of fair? The first six chapters provide a leisurely survey of the problem, chapter 7 is a quick reference which summarizes known results, while the last four chapters contain technical proofs for previously introduced problems.

The book contains only "elementary" results in the sense that the work is almost completely self-contained, even for the novice. The algorithms we present rarely apply complex theorems from other areas in mathematics, although there are some exceptions. Also, we have consciously avoided using the most abstract setting and weakest possible hypotheses to gain greater generality.

With its excursions in many directions, this book can serve as a text for a course. Exercises are included which serve the varied purposes of providing routine extensions of material covered for the novice, checking details omitted in the exposition, or suggesting directions for new and substantive exploration. The book can also serve as a supplement or source for special projects in mathematics courses for liberal arts students; modeling courses; and courses in number theory, graph theory, complexity or general combinatorics. It provides a reference for quick access to currently known algorithms for various cake-cutting problems.

We owe our own personal introduction to this subject to Professor William Lucas, to whom we express our gratitude. Conversations with Steven Brams, Alan Taylor, Theodore Hill, Kevin McAvaney and others have helped to focus our interest. The Combinatorial Geometry Seminar at Washington State University has played the role of a long-suffering sounding board for our ideas over a period of years.

We also wish to thank John Burke and Kenneth Davis for looking over earlier versions of the manuscript. Special thanks go to Jaimie Dahl for her endless patience in typing the manuscript as we changed our minds about what we had written.

We appreciate the interest of all readers and trust that some of our own enthusiasm for this beautiful subject can be shared.

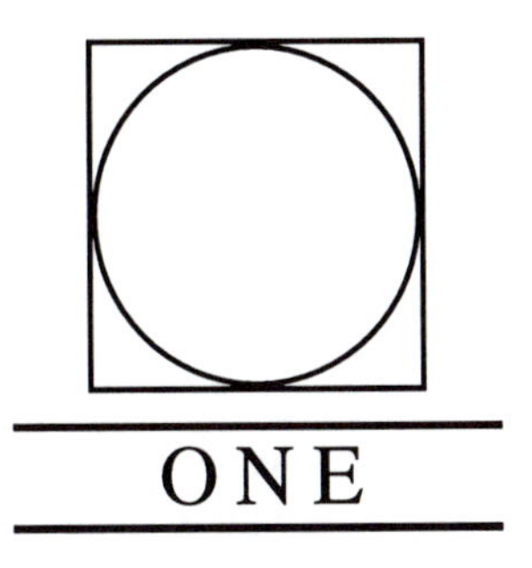

ONE

Fairly Dividing a Cake

1.1 First Things

Once upon a time (in fact, many times down through the ages) there was a leftover morsel to be divided "fairly" among two or more persons. How should this be done? Therein lies the mathematical tale of this book. What does "fairly" mean and how can the required division be accomplished?

Issues of fairness surround us in our daily affairs. Are we fairly taxed? Did the judge make a fair decision? Are prices fairly set? Were the playground teams fairly chosen? Was a referee's call a fair call? Is this nuclear weapon treaty fair to our country? Were congressional districts fairly drawn? These issues range in importance from the frivolous to those over which people fight and die.

"Fairness" is a difficult thing to judge, partly because people don't agree on what "fair" means. We can get two strongly different opinions about whether the referee made the correct call, depending on which of two people we ask. Can "fairness" be quantified in some way and thus be studied with dispassionate mathematical tools, or is the issue too subjective to permit a mathematical handle?

The objective of this book is to examine one setting, fairly dividing a cake (or any other item which can be cut into pieces without destroying its value), in which fairness can, in some sense, be measured and thus studied mathematically. If this problem is unfamiliar to you, your first reaction may be: "You mean a whole book on cutting a piece of cake? How can there be that much to say on the subject?" Besides being inherently interesting, the subject is surprisingly rich mathematically. A considerable literature on this clearly focused problem has developed over the past fifty years, much of it very recently. This circumstance justifies gathering a good portion of those scattered works into one volume.

In a broader perspective, the mathematical studies of this book can be viewed as part of the ever expanding role of mathematics in addressing problems from the social sciences. Many feel that mathematics will receive as much of its direction and vigor in the future from problems in the social sciences as it has received in the past from the physical sciences. Many of these problems spring from issues of fairness, and there are some important and recent mathematical results which give new insights into age-old difficult social problems. Arrow's Impossibility Theorem [Bla] about fair voting procedures and the work of Balinski and Young [BY] about fair methods of apportioning representative bodies are such examples. Even on the specific topic of fair division it is possible to take a wider view of the subject, as in the recent work of Brams and Taylor [BT3]. Our focus, however, will remain primarily on algorithms for cake-cutting.

It is safe to say already that the use of mathematical applications in the social sciences is rapidly expanding. One fortunate aspect of this expansion is the turning of new, fertile soil where fresh problems abound for mathematical consideration. For example, one does not have to ponder the work on cake-cutting for long in order to formulate a number of interesting and new questions. As always, some are too hard to make quick progress on, but happily a number do submit to reasonable mathematical effort. This book represents a catalog of examples of such successes. Also, many new questions will suggest themselves along the way.

The purpose of this introductory chapter is to state the cake-cutting problem in some of its various renditions and examine some solutions which have been given. We will avoid unnecessary notation or technical paraphernalia which might detract from what we feel is the simple beauty of the problem. We encourage the reader to envision yourself in a life-or-death struggle with siblings to see that you get a "fair share" of a literal leftover piece of your favorite cake. Don't give up a crumb!

The problem traces its beginnings to the distinguished group of Polish mathematicians whose names appear on numerous theorems of the mid-twentieth century, particularly theorems in analysis. The problem was presented by Hugo Steinhaus to the mathematical and social science communities on September 17, 1947, at a meeting of the Econometric Society in Washington, D.C. (although private correspondence between Steinhaus and Bronislaw Knaster is reported in 1944 [Kna]). In a session chaired by Herman O. A. Wold of Sweden, Steinhaus, whose name graces a number of important theorems in mathematical analysis, stated two specific problems of fair division and gave algorithms which guaranteed fairness in those two settings. A report on the session notes that "Mr. Steinhaus' paper was discussed by Mr. Tjalling C. Koopmans and the speaker." The solutions which Steinhaus gave were attributed to Knaster and Stephan Banach [Ste1]. Recent years have seen a heightened interest in the problem with many resulting articles. Thus, knowing that the problem has a fifty year pedigree with impressive ancestry, let us return to that piece of cake which must be divided in a satisfactory way between two hungry, competitive siblings.

One critical feature which makes the problem rich is that these two people will disagree on the value of a piece of cake. How can this disagreement be accommodated in our study and how will it affect our basic understanding of fairness? Should mother cut "fair" pieces and assign them to the two hungry children? What may look like fair halves to her may not look like halves at all to the children. Shall we get out a scale and weigh them? Merely weighing out two equal pieces and assigning them to the two children may well produce dissatisfaction. The cake may be non-homogeneous (one chocolate layer; one white layer; mocha frosting here and cream filling there), and so the two may disagree about how much a certain piece is worth. If Tom likes chocolate, he won't be happy if the piece assigned to him is all or mostly white cake — even if it is the same weight as the other piece that is mostly chocolate. If Dick also wants the chocolate, they could have a serious problem. Of course, if Dick prefers the white cake, they both could go away satisfied. How should we accommodate the disagreement and to what extent is it an intolerable mathematical nuisance?

We will see in a number of places that disagreement on the value of a piece can actually produce better results than we would get if there were total agreement. In the words of Steinhaus, "It may be stated incidentally that if there are two (or more) partners with different estimations, there exists a division giving to everybody more than his due part; the fact disproves the common opinion that differences in estimations make fair division difficult." More on this theme follows in Chapter 4.

The place to start any discussion of fair division is the ancient method, which Brams and Taylor [BT3] have traced back twenty-eight hundred years, of "one cuts; the other chooses" or "one pours; the other picks." However, some of the most important principles of cake-cutting can be best illustrated by inadequate methods. So let's examine Cut and Choose and three other methods we might consider using to divide the cake.

We can immediately see that some methods are fairer than others. Are any of the methods completely satisfactory and on what basis? Using Method 1, we can trust Mom to cut what she considers equal halves, but neither Tom nor Dick may receive a piece he considers half of the cake. Tom likes plain cake, Dick likes icing, and Mom is indifferent about the two. Hence the pieces may be

Method 1. Mom cuts the cake in two pieces (which we assume she judges equal) and assigns the pieces to Tom and Dick.

Method 2. Mom cuts the cake in two pieces; Tom and Dick flip a coin to see who chooses first with the other claiming the remaining piece.

Method 3. Tom cuts two pieces and Tom gets first choice.

Method 4. Tom cuts two pieces and Dick gets first choice.

valued differently by all three. Mom says they are 50% – 50%; Tom sees them as 40% – 60%; and Dick sees them as 70% – 30% in reverse preference to Tom. Note that while each has assigned values that add to 100%, if Mom assigns pieces one way, Tom and Dick think they have portions representing only 40% and 30% of the cake respectively, even though all of the cake is distributed. In this case, the problem is easily resolved by having them trade pieces. Then Tom feels he has 60% of the cake while Dick feels he has 70% of the cake.

Further trouble is seen with Method 1. What if Tom and Dick both judge the same piece as larger? Then one is sure to feel he did not receive a fair half and nothing can be done about it. Without allowing a switch of pieces (which the method description should explicitly mention if it is allowed), Method 1 does not guarantee even one sibling a fair half by his own assessment. Even with switching we can not guarantee both a fair piece, so Method 1 does not work.

Now, Method 2 will surely give the person who wins the coin flip a piece that person considers at least a half, and maybe more, because that person chooses first. (If two non-negative numbers sum to one, one of the numbers must be at least one-half.) And since either is equally likely to win the flip, there is no favoritism toward either player. But if we ask the specific question, "Is Tom guaranteed a half of the cake by Method 2?," the answer is "No" because he may lose the coin flip. Does the method guarantee him at least 10% of the cake? Again the answer is "No" because Mom's two pieces may look in the ratio 5% – 95% to Tom; in this case Tom may lose the toss and be left with the piece he considers 5% of the cake. So Method 2 does not guarantee Tom (or Dick) any fixed amount of cake. Tom's problem is in the method and not that either Mom or Dick is out to cheat him or that they are in some sort of sinister collusion. The problem has occurred because benevolent people have honest differences of opinion about what pieces of cake are worth. Moreover, in less friendly company we shouldn't assume that everyone will act benevolently.

We can be sure Dick will object to Method 3, because there is nothing to prevent Tom from cutting highly unfair pieces since he gets first choice. There is the remote chance, since he is required to cut two pieces, that even though he purposely cuts himself a very large piece, Dick will see the other piece as at least a half. But that would be coincidental, based on two different views of what pieces of cake are worth, and not a situation the method would always guarantee. Dick shouldn't have to pin his hopes on lucky disagreement to get his fair share.

Maybe we are now ready to be a bit more explicit about what we are trying to accomplish when we ask for a "fair" division. Can we find a method of division that will guarantee both Tom and Dick at least a half of the cake *by their own assessments?* This phrase, "by their own assessments," identifies the central issue which must be addressed in any fair division method. None of the first three methods do that; in fact, only Tom in Method 3 is guaranteed such a piece. But look at Method 4. We assume Tom can and will cut the cake in two pieces about which he is indifferent (i.e., 50% – 50%) since he knows he may get either piece. So he is guaranteed exactly 50% of the cake if he cuts exact halves. If he doesn't cut halves, he could get a smaller piece, but that is a problem with his strategy or cut, and not a problem in the method. Since Dick chooses first, he is guaranteed at least half of the cake. Method 4 works!

We now formally describe this successful Cut and Choose method which we will see again many times.

Cut and Choose Algorithm

Step 1. Either person cuts what he or she considers equal halves.

Step 2. The other person chooses; the remaining piece goes to the cutter.

From this very simple and elegant beginning, a large number of interesting variations on the theme are suggested. What if there are more persons or players among whom to divide the cake "fairly"? Are there other definitions of "fair"? What if the cake is to be divided in unequal shares? Maybe yesterday Tom got a larger piece, so that today's division should be in the ratio of 40% for Tom and 60% for Dick. There is plenty to be explored, so let us take a reasonable next bite of the problem.

1.2 On to Three Persons — A False Start

Having satisfied Tom and Dick, we now ask what would happen if Harry joins the proceedings. That is, can we devise a method which will give each of three persons or players a piece of cake that he or she considers to be at least one-third of the cake? We will call this problem *simple fair division.*

This jump from two to three persons presents an interesting and challenging question. To again illustrate the flavor of the problem we will try a method which doesn't work and find out why it doesn't. Let us call the cake to be divided X, and use some familiar notation to simplify things.

Doesn't Cut It Method

Step 1. Tom cuts X into two pieces, $X = X_1 \cup X_2$, so that he thinks X_1 is 1/3 of X and X_2 is 2/3 of X.

Step 2. Dick cuts X_2 into two pieces, $X_2 = X_{21} \cup X_{22}$, which he considers halves of X_2.

Step 3. Harry chooses one of the pieces X_1, X_{21}, or X_{22}; then Tom chooses one of the two remaining pieces; Dick gets the last piece.

Who is always satisfied by this method? Well, certainly Harry is, because he gets first choice of the three pieces, so at least one of them must look like at least 1/3 of the cake to him. Whichever it is, he can have it.

Let us next show that Tom also is guaranteed at least 1/3 by his assessment. To demonstrate this, let's see how many of the three pieces X_1, X_{21}, and X_{22} Tom finds acceptable, i.e., how many contain at least 1/3 of the cake by his assessment. Certainly X_1 is acceptable because he was told to (and presumably did) cut X_1 so that it was an exact third of X. Since Dick cut X_{21} and X_{22}, we don't know what Tom thinks of them. But we do know that X_2 was 2/3 of the cake in Tom's eyes, so at least one of X_{21} or X_{22} must be worth at least 1/3 to Tom. Thus, we know that Tom will accept at least two of the three pieces, and as second chooser he can claim one of these two acceptable pieces. So far so good!

Before moving on to Dick, let us observe that if Tom cut anything other than 1/3 on his cut, he might not receive a full third by his assessment. If Tom cuts X_1 less than 1/3, Dick may cut X_2 in two pieces only one of which Tom would consider fair. Because he chooses second, that piece may be gone when he chooses. On the other hand, if he cuts X_1 larger than 1/3, then X_2 is less than 2/3 and Dick may cut X_2 in exact halves, not only by his, but also by Tom's assessment. Then only X_1 would be acceptable to Tom, and Harry may choose X_1. So we see that if Tom is to be guaranteed at least 1/3 of the cake he must cut exactly 1/3 as instructed.

How about Dick? How many pieces are we sure he would accept as fair? If he agrees with Tom's first cut, then all three will be acceptable to him (why?) and all is well. But we have no guarantee that he agrees with Tom. If Dick thinks X_1 is 70% of the cake, which might be the case even though Tom thinks it is only a third, then neither X_{21} nor X_{22} is acceptable. To be satisfied Dick would have to get X_1, but in choosing third there is no guarantee that X_1 will be left for him. Hence our method fails to always satisfy Dick. In fact, we can quickly see that the only time Dick is guaranteed 1/3 of the cake using this method is when he agrees with Tom's cut. (See Exercise 1.2.) In the rest of this introductory chapter we will see some algorithms which do guarantee all three people at least a third of the cake by their assessments. Before you read on, we encourage you to try to devise a method of your own.

1.3 On to Three Persons — Successfully

There is a simple and elegant way to give fair shares using a continuously moving knife which was described by Dubins and Spanier [DS].

The Moving Knife

Think of the cake sitting on a table and a knife held above the cake ready to make a cut. Begin with the knife completely to the left of the cake, and gradually and continuously move the knife to the right, passing over the top of the cake. Let X_1 be the piece to the left of the knife and X_2 the piece to the right. As the knife moves from left to right, instruct Tom, Dick and Harry to say "Stop" when the value of X_1 first becomes 1/3. When the first player says "Stop," cut the cake at this position and give piece X_1 to that player who then drops out of the proceedings. Note that each of the other two players thinks that X_2 is at least 2/3 of the cake. Now continue to pass the knife across the top of the piece X_2, creating pieces X_{21} to the left of the knife and X_{22} to the right. When the first remaining player to think X_{21} is a third of the cake says "Stop", give X_{21} to that player and X_{22} to the other player.

Why are all three satisfied with this procedure? From the instructions given, we know that both players who say "Stop" are given a piece valued at exactly 1/3. The player who never says "Stop" has given away two pieces, neither of which he considers 1/3, so at least 1/3 is left for him. Think about what should be done if two players say "Stop" simultaneously.

When performing "Moving Knife" how can we be sure the first person to say "Stop," let's assume Dick, did so when he felt exactly 1/3 of the cake had been swept over? We can't, because we don't know how he values the cake. Maybe Dick has actually held out for 40% of the cake. What is to prevent that? Only the risk of his not getting a full third, for if while holding out for more, another player says "Stop" when Dick thinks 39% of the cake has passed under the knife, he will then have to do the best he can in dividing what he considers as 61% of the cake with another person. He may then not get his 1/3.

The extension of the Moving Knife Algorithm to more than three players is now formally stated, and you can easily extend our argument from three players to n players to see that all players are satisfied.

Moving Knife Algorithm for Simple Fair Division

Step 1.	A knife is continuously passed over the cake from left to right. The first player who thinks the portion to the left of the knife is $1/n$ of the whole cake says "Stop." That piece is cut and given to the player who said "Stop" and that player drops out. If two or more players say "Stop" simultaneously, any one of these players can be assigned the piece.
Steps 2 through $(n-1)$.	Repeat Step 1 with the remaining players on the remaining piece of cake.
Step n.	There is now one player left. Give that player the remaining piece.

When only two players remain, should one say "Stop" when the piece to the left of the knife is $1/n$ of the cake, or ignore the instructions and wait until the pieces on either side are equal? Either will guarantee that all players receive a fair piece. If you happened to be one of the last two, would you prefer to say "Stop" when $1/n$ would be cut or when the last two pieces are of equal size? If you think more than $2/n$ remains, you can hold out for half of what is left (rather than settling for an exact $1/n$) at no risk. This is one of those few exceptions where not following the exact instructions given in the algorithm will not expose the player to the risk of losing a fair share of the cake. Do you think any other player could ignore the instructions without risk?

In one form or another, what we have just seen appears in many cake-cutting algorithms: namely, can we somehow cut a piece that some of the players are willing to share and the rest are willing to give away while taking their portion from the other piece. Watch for this in future algorithms. In the rest of the book many algorithms for fair division will be presented. In the instructions, values of cake to be cut will be given which, if followed, will assure the required outcome for all players. If the person being instructed chooses to cut otherwise, that player will generally (but not always) risk not getting a fair share. It will always be the case that, should a person choose to take such a gamble (for example, Dick holding out for 40% above), no other players' fair share will be compromised.

You may have detected a substantial difference in the Moving Knife Algorithm and all the others we have considered. For example, in Cut and Choose for two players, only two decisions are required. One person has to decide what a half is and cut it, the other has to evaluate the two pieces produced and choose the larger. Only a finite number of decisions are made. For that reason, it is called a *finite* algorithm. (This is discussed more fully in Section 2.1.)

In the Moving Knife Algorithm, however, each player must make a yes-no decision at each instant of time: namely, is the piece of the cake to the left of the knife worth $1/n$ or not? Because players have to make decisions at each point in a continuum of knife positions, the Moving Knife Algorithm will be called a *continuous* algorithm. Continuous algorithms tend to be more powerful than ones where only a finite number of decisions are made, such as Cut and Choose. For example, Moving Knife accomplishes fair division for three players with two cuts and three pieces. Later we will see that you cannot accomplish fair division for three persons with two cuts (three pieces) using *any* finite algorithm.

The Trimming Algorithm

The next algorithm we look at is essentially a modified version of Moving Knife (although historically it preceded the Moving Knife). The objective is to identify where the first person would have said "Stop" without moving a knife. This algorithm, which we will call the "Trimming Algorithm," is due to Banach and Knaster [Ste1].

Let us first ask Tom to cut off what he considers to be 1/3 of the cake. Pass that piece to Dick and ask him to trim it back to 1/3 if he considers it to be more than 1/3, otherwise he leaves the piece alone. Dick passes the resulting piece (trimmed or untrimmed) to Harry who can either take it or leave it. If he chooses to take it because it appears to be at least 1/3 of the cake to him, he is satisfied, drops out, and Tom and Dick play "Cut and Choose" on what is left. (If any "trimmings" result, those bits become part of what is left.) If Harry does not accept the piece and Dick has trimmed it, then Dick takes it; otherwise Tom gets it. The remaining two players play "Cut and Choose" on what is left. Note that in all cases either Harry or the last person to cut gets the first piece.

Now let's check that Tom, Dick, and Harry all agree that they each have received at least 1/3 of the cake. The person who takes the first piece thinks that piece is at least 1/3 of the cake. Check this by considering cases of who gets the first piece (and it could be any one of the three). Also, the other two agree that at most 1/3 is being given away because the first piece to be given away does not pass by Tom or Dick in excess of 1/3 of the cake by their estimations. Since the two remaining players agree that at most 1/3 was given away, then they also agree that at least 2/3 remains to be divided. Finally, "Cut and Choose" guarantees each of the last two players at least 1/2 of what is left, and 1/2 of

2/3 is 1/3 as required. Let us now formally state the trimming algorithm for n players. You should convince yourself by generalizing the argument above that each player is guaranteed at least $1/n$ of the cake by his or her estimation.

Trimming Algorithm for Simple Fair Divison

Step 1. Player P_1 cuts a piece of size $1/n$ from the cake.

Step 2. The cut piece is passed successively to $P_2, P_3, \cdots, P_{n-1}$. Any player who thinks the piece he or she is passed exceeds $1/n$ in value trims it so the reduced value is exactly $1/n$.

Step 3. Player P_n takes the trimmed piece resulting from Step 2 if P_n considers it at least $1/n$ of the cake. Otherwise the trimmed piece is given to the last player who trimmed it. The player receiving this piece drops out.

Step 4. Repeat Steps 1–3 on the remaining portion of the cake with n replaced by $n-1$ and the players renamed P_1 to P_{n-1}. Repeat this step until one player remains.

Note that this algorithm is just Cut and Choose when $n = 2$. Next we will look at a completely different way to satisfy Tom, Dick, and Harry. We will call this new method the "Successive Pairs Algorithm" because it will be a series of divisions between only pairs of players [Saa].

The Successive Pairs Algorithm

Since "Cut and Choose" solves the problem for two players, maybe we could extend that idea to three players. If we know how to solve the problem for n players and we are asked to solve it for $n + 1$ players, it is always a good idea to see if we can utilize the solution for the n players in the solution for $n + 1$. This is the essence of mathematical induction.

Suppose Tom is by himself and sits down to enjoy the entire piece of cake. At that moment Dick shows up and demands a share. So, let Tom cut halves and Dick choose a piece. Just as Tom and Dick prepare to enjoy their cake, along comes Harry who demands that he be given a fair third. Tom and Dick now each have at least 1/2 of the cake but are supposed to get only 1/3. So each needs to give up some of what he has to Harry. How can this be done so that Harry will get a combined share he considers 1/3 of the entire cake? If Tom and Dick each cut their piece in what they consider exact thirds, each piece (to them) would be at least 1/3 of 1/2, or 1/6. Have each of them do that and let Harry take one of the resulting three pieces from both Tom and Dick. Is everybody satisfied?

We know Tom gets exactly 2/3 of exactly 1/2 by his own assessment. We also know Dick gets exactly 2/3 of at least 1/2 by his own assessment. (Note the change of "exactly" to "at least.") So they are both satisfied. But what about Harry? He gets at least 1/3 of everything. More formally, if Harry thinks Tom's piece is worth α, where $0 < \alpha < 1$, then he thinks Dick's piece is worth $(1-\alpha)$. Thus, Harry gets at least $\alpha/3$ from Tom and at least $(1-\alpha)/3$ from Dick. But $\alpha/3 + (1-\alpha)/3 = 1/3$, so Harry gets at least 1/3.

Now we will formally state the Successive Pairs Algorithm for n players. Note in this case we will work "upward," doing the procedure inductively with increasing n. On the other hand, the Trimming Algorithm works "downward" inductively for decreasing n.

Successive Pairs Algorithm for Simple Fair Division

For $n = 2$ the algorithm is the Cut and Choose Algorithm. To get from $n-1$ players to n players, $n = 3, 4, 5, \cdots$:

Step 1. Have players $P_1, P_2, \cdots, P_{n-1}$ perform the algorithm for the case $(n-1)$. (So each thinks he or she has at least $1/(n-1)$ of the cake, composed of a number of pieces.)

Step 2. Each of $P_1, \cdots, P_{n-1}$ cuts each of his or her pieces into n equal parts.

Step 3. Player P_n chooses one part from each piece cut in Step 2.

Notice that each of $P_1, \cdots, P_{n-1}$ gets exactly $(n-1)/n$ of at least $1/(n-1)$ of the cake, that is, at least $1/n$. On the other hand, if P_n values the entire holdings of $P_1, \cdots, P_{n-1}$ to be $\alpha_1, \cdots, \alpha_{n-1}$ respectively (so that $\alpha_1 + \cdots + \alpha_{n-1} = 1$), then P_n gets at least $(1/n)(\alpha_1 + \cdots + \alpha_{n-1}) = 1/n$ also. You may have noticed that this method generates a lot of pieces. We will return to this issue in Chapter 2 where we will revise the method so that fewer cuts are needed.

1.4 Envy Rears Its Ugly Head

We have seen various ways to guarantee Tom, Dick, and Harry all at least a third of the cake by their own assessments. In that sense they have been fairly treated. Nevertheless, it may be the case that, while Tom has his third, he still feels that Dick got a piece bigger than his. It is one thing for Tom to get a fair piece (at least a third) but quite another for him to get the largest piece. If he thinks Dick has received a larger piece, it is human nature that Tom is going to have second thoughts about the "fairness" of his share.

This simple observation gives us another way we might interpret "fair" and another task of trying to find ways to divide the cake to satisfy this kind of fairness.

A cake division is envy-free if no player feels another has a strictly larger piece.

How can we accomplish envy-free division? Are any of our previous algorithms envy-free? Note that while guaranteeing each player $1/n$ of the cake does not mean the division was envy-free, any envy-free division must give each player at least $1/n$ of the cake. If I have been given less than $1/n$, I am sure to feel somebody has a bigger piece than mine or else the sum of the parts wouldn't equal one. This all means we should expect envy-free division to be a harder problem than just simple fair divison. In this expectation we will not be disappointed.

As a first step, let's review some of our earlier algorithms to see whether or not they are envy-free. Certainly Cut and Choose for two players is envy-free (why?). As for Moving Knife, nobody will envy the piece that the first person to say "stop" gets, since nobody else thinks the piece has size $1/n$ and they are guaranteed to get at least that much later. One can pursue this reasoning to see that, for example, the tenth person to receive a piece will not envy any piece that the nine previous players have received. On the other hand, the first player to get a piece might envy some (but not all) of the later players.

The fact is that none of our earlier algorithms, other than two person Cut and Choose, is envy-free. This will be pursued later in the exercises. This leaves us the task of trying to produce a new algorithm which will do the job. Even for three persons, it is not an easy task.

The following method for finding an envy-free division for three players is one of the prettiest in the subject of cake cutting. The algorithm is attributed to John Conway, Richard Guy, and John Selfridge in [Str] and to Selfridge in [Wool]. Some of the ideas which appear here will also be used in some of the more general envy-free algorithms we will look at later.

Let's begin with an informal description to motivate some important ideas. Begin by having Tom cut the whole cake X into three equal pieces X_1, X_2, X_3. No matter who gets these pieces, Tom will be envy-free because he thinks they are all equal. What about Dick and Harry? Maybe Dick thinks X_1 is the largest and Harry thinks X_2 is the largest (we should be so lucky!). Clearly giving X_1 to Dick, X_2 to Harry and X_3 to Tom would make the divison envy-free.

But what if both Dick and Harry think X_1 is the largest? Whichever one doesn't get X_1 will envy the person who does get it. Notice that Tom is envy-free because he was the one who cut the pieces to be equal. Can we create pieces so that Dick is likewise envy-free?

We can suppose that Dick orders the pieces X_1, X_2, X_3 from largest to smallest. What if Dick trims off an excess piece E from X_1 so that he thinks $X_1' = X_1 - E$ is equal in size to the second largest piece X_2? If Dick already thought X_1 and X_2 were equal, then no trimming would be needed.

This idea of trimming so that Dick believes X_1' and X_2 are equal and largest is the first of the key ideas in this algorithm. Note that for both Tom and Dick there is more than one piece tied for the biggest. Now Harry can have his choice among the three pieces X_1', X_2, X_3, thereby guaranteeing him a preferred piece. Next, Dick can have whichever of X_1' or X_2 is left, also guaranteeing him a preferred piece. If both are left after Harry picks, we will insist that Dick take X_1'. Now Tom gets the remaining piece, which must be either X_2 or X_3, so he also gets a preferred piece. Since all three received a preferred piece, they are envy-free so far.

But what about the left-over piece E? It looks like we still have to divide it in an envy-free manner. If we just repeat the steps given above on E, we will produce yet another left-over piece. Clearly we're in danger of an infinite process.

But wait a minute. We know something about E that wasn't true of the whole cake. Since Tom received either X_2 or X_3 and he originally cut X_1 to be equal to these two pieces, he would remain envy-free of the person getting X_1' **even if that person additionally got all or any part of E**. This is the second key idea in this algorithm.

Suppose Dick got X_1'. Have Harry cut E into three equal pieces. Then let Dick choose first, Tom second, and Harry third. Dick gets a preferred piece of E. Tom doesn't care how much of E Dick got (by the observation above), and he picks before Harry. Harry cut the pieces to be equal, so he remains envy-free. We've done it! (What if Harry got X_1? Only a minor modification is needed.)

Envy-Free Algorithm for Three Players

Step 1. Have Tom cut three equal pieces which Dick ranks X_1, X_2, X_3 from largest to smallest.

Step 2. Have Dick trim off E from X_1 (if necessary) so that $X_1' = X_1 - E$ and X_2 have equal size.

Step 3. From among X_1', X_2 and X_3 choose in the order Harry, Dick, and Tom. If available, Dick must choose X_1'.

Step 4. Either Dick or Harry received X_1', call him person P_1 and the other person P_2.

Step 5. Have P_2 cut E into three equal pieces which are chosen in the order P_1, Tom, P_2.

This clever algorithm does a nice job for three people. Can we do envy-free division for four or more people? The answer will turn out to be yes, but the algorithms seem to get much more complicated.

At this point, we have experienced some of the flavor of fair division algorithms. We have also seen, just by cracking the door on the problem, some of the directions in which our problem may lead us. We need to look at other interpretations of "fair." Like all mathematicians, we want to be efficient and accomplish our divisions with as few steps as possible. We also have various classes of algorithms to use, and it will be interesting to note what we can accomplish within each class.

1.5 EXERCISES

1.1. In the Cut and Choose Algorithm, would you prefer to be cutter or chooser or does it matter? Why?

1.2. Show that the only time Dick is guaranteed at least 1/3 of the cake using the Doesn't Cut It Method is when he agrees with Tom's cut.

1.3. In the Trimming Algorithm for three persons where Tom makes the first cut and Dick then trims, would you prefer to be Tom, Dick, or Harry? Why?

1.4. In the Successive Pairs Algorithm:

(a) Would you prefer to be Tom, Dick, or Harry? Why?

(b) Describe a way Harry could cut pieces held by Tom and Dick and still guarantee all 1/3 of the cake. Would Harry prefer this method or the original?

1.5. Compare the number of cuts in the worst case scenario for the Trimming and the Successive Pairs Algorithms for three players; for four players.

1.6. Who among Phil, Jill, and Will is guaranteed 1/3 by each of the following algorithms?

(a) Phil cuts halves and Jill chooses. Then each cuts the other's piece into what the cutter considers equal thirds. Finally Will takes a piece from each player.

(b) Jill cuts equal thirds, and then Will cuts each of those pieces in halves. The six pieces are chosen in the order PWJJWP.

(c) Cut as in (b) but choose in the order PJWWJP.

1.7. Devise an algorithm in which none of the three players is guaranteed a third.

1.8. This exercise presents yet another valid algorithm described by Kuhn [Kuh] for fairly dividing the cake. Have Tom cut what he considers equal thirds and ask the other two to identify any of the three pieces they find acceptable (worth a third in the player's estimation). Let us organize this information in a matrix where "1" means "this piece is acceptable" and "0" means "this piece is unacceptable." The information might look like the table below, for example.

	X_1	X_2	X_3
Tom	1	1	1
Dick	1	1	0
Harry	0	1	0

(a) How can the division be accomplished in this case?

(b) Why may we assume that all entries in the first row are "1"? Can there be a row with no "1"?

(c) Devise a method of fair division if the table looks like:

	X_1	X_2	X_3
Tom	1	1	1
Dick	1	0	0
Harry	1	0	0

(Of course, further cuts will be required.)

(d) Considering the problem for four players, how could you proceed if the matrix took the form given?

	X_1	X_2	X_3	X_4
Tom	1	1	1	1
Dick	1	0	0	0
Harry	1	0	0	0
Amy	1	0	0	0

(e) Assuming that there is always a fair division for three players, consider the problem for four players and show in all cases that a fair division is possible.

Comment: The extension of this method to n players is more involved than the ones we have seen in Chapter 1. As you would expect, it relies on a combinatorial result about matrices with entries of 0 and 1. The details can be found in [Kuh]. (A modified version using graph theory is found later in Chapter 6.)

1.9. In the envy-free algorithm, who among Tom, Dick, and Harry is guaranteed 1/3 of the cake before E is divided? Is this division of E envy-free?

1.10. (a) Suppose the cake has been cut $X = X_1 \cup \cdots \cup X_n$ so that player P_i thinks there are exactly i acceptable pieces, $i = 1, 2, \cdots, n$. Is fair division always possible in this case?

(b) If all n players in (a) had been asked piece by piece if it was acceptable or not, of the n^2 questions asked, the answer would have been "yes" in $1+2+3+\cdots+n = (n^2+n)/2$ times. Give a case where there are that many "yes" answers but no fair division is possible without further cuts.

(c) What is the fewest number of "yes" answers which guarantees that a fair division is possible?

1.11. Suppose the cake is cut in six pieces and you get first and last choice.

(a) How much are you sure to get by your estimation on your first choice? The last choice?

(b) Show the two combined choices will always guarantee you at least 1/5 of the cake. Under what conditions will you get only 1/5 of the cake?

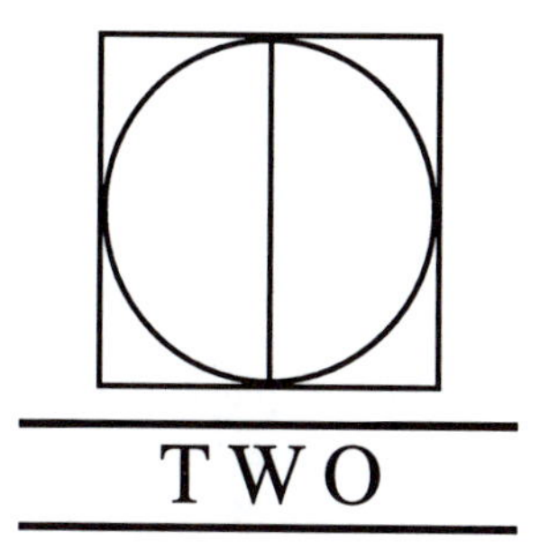

TWO

Pieces or Crumbs — How Many Cuts Are Needed?

2.1 First Things — Some Agreements on Assumptions and Notation

When introducing the cake-cutting problem, Steinhaus raised the issue of the number of cuts used for fair division with the comment: "Interesting mathematical problems arise if we are to determine the minimal number of cuts necessary for fair division" [Ste1].

Certainly if you are actually dividing a cake, you don't want to receive a pile of crumbs for your fair share. You prefer not to make too many cuts. More realistically, the issue at hand is a mathematical one. The general problem of making algorithms as efficient as possible is important, and its solution can save large amounts of money in computer time in certain settings. Indeed, it may be the case that, although an algorithm is known to accomplish some task and computers can be programmed to do the work, there are so many steps required that even today's computers cannot give an answer in your lifetime. Besides these practical considerations, there is always the esthetic motivation of doing something in as efficient a way as possible. Discovering how many steps are required by an algorithm is an issue addressed in the relatively new area of computational complexity.

As Steinhaus has said, "interesting mathematical problems" lie ahead as we consider the number of cuts required for fair division. But we know that the

Moving Knife Algorithm will guarantee each of n players at least $1/n$ of the cake using $(n-1)$ cuts to create n pieces, and we certainly can't achieve our goal with fewer than $(n-1)$ cuts. Still, Steinhaus raised the question of the number of cuts required for fair divison. So where is the issue?

It is clear that Steinhaus was thinking only of *finite* algorithms, ones where a finite number of discrete steps are used. On the other hand, as we have noted, the Moving Knife Algorithm requires players to make continuous decisions as time proceeds, so any moving knife algorithm (and we will see others) is not permitted in the contest for fewest number of cuts.

Methods for dividing a cake fairly have variously been called algorithms, protocols, procedures, etc. Various definitions of these terms have been given (see for example [BT2], [EP]), but for the most part we will take a more informal approach to the algorithms, the term we will use. Nevertheless, we do need to agree on some things we assume about algorithms.

An algorithm will be a clearly stated sequence of instructions, designed to accomplish a specific task, such as "Give each of three players a piece of cake that each considers at least one-third of the whole." One important feature of our algorithms will be the inclusion of a strategy each player can follow to guarantee the desired result. For example, in Cut and Choose, the first instruction is for the cutter to "cut halves" rather than to "cut two pieces." The only strategy the cutter can follow to guarantee that he or she actually receives a half is to cut halves. If other than halves are cut, the cutter may receive the smaller piece.

The cake-division problem can be thought of as a game where each person is out to get the most cake possible. As William Poundstone has noted regarding the two person Cut and Choose Algorithm, "This homely example is not only a game in von Neumann's sense, but it is also about the simplest illustration of the minimax principle upon which game theory is based" [Pou].

The piece to be cut at a given stage may be determined by the answer to questions such as: "Which piece do you consider largest?" or "Do you think this piece is worth at least 1/4 of the whole cake?" Based on such information, which may include how each player judges existing pieces, the algorithm will direct a player to make a next cut of specified size. If the instructions of the algorithm are followed, the desired fair division will be accomplished.

In a sense, an algorithm is somewhat like a recipe. If the goal is to bake a tasty chocolate cake, the recipe specifies ingredients, order of mixing, temperatures for the oven, and baking time. One step might be "add a cup of sugar." Unlike algorithms in which we may require a specific player to make the next cut, the recipe does not care who adds the sugar. Let us suppose Tom is to do it and he decides to add salt instead. His decision to ignore the instructions will assure that the cake is spoiled for everyone.

Likewise in cake-cutting algorithms, Tom may decide not to cut halves even though the algorithm tells him to. Generally he does so at some risk, but always only to himself. Adding salt ruins the cake for everyone, but not following the directions of an algorithm does not put the fair share for other players at risk.

In fact there is no real way to police the behavior of the players. If Tom is told to cut halves and he makes a cut, does he really cut halves? If he expresses an opinion about a piece in response to a question, does he tell the truth? Only Tom knows, because nobody else knows what Tom actually thinks of any piece. We have only Tom's word for it.

When the referee or algorithm tells a player to cut a certain piece in half, for example, this can be interpreted as meaning:

- The player must cut the indicated piece but can choose to cut in anyway he or she wants.

- If the player cuts the piece in half, then the player is guaranteed an acceptable piece. If any other strategy is used, the player may or may not receive an acceptable piece.

Thus, our descriptions of algorithms combine the precise directions concerning which piece to cut, along with one possible winning strategy for cutting the pieces in order to guarantee fairness.

Before preceeding to count the pieces required for fair division, some additional observations are needed. We have already utilized a number of unstated assumptions. Let us explicitly list some of them and agree on some notation that will help us present our thoughts.

- All players are entitled to their own opinions about a piece of cake. We will usually write X for the whole cake that is to be divided. Subsets (or pieces) of the whole will be denoted by letters $A, B, C, \cdots$ or $X_1, X_2, \cdots$. Since Tom, Dick, and Harry are all entitled to their own opinions about a piece A, let us write $\mu_T(A), \mu_D(A)$ and $\mu_H(A)$ for those three values. The letter μ is associated with functions, called measures, with certain properties we will soon state, and the subscript T identifies that function with Tom. When the players are $P_1, P_2, \cdots, P_n$ we will write $\mu_i(A)$ for the value player P_i places on piece A. In this way, each player has his or her own function with values on the pieces independent of anyone else's values. We do insist that they agree the whole cake X has value 1, so that $\mu(X) = 1$ for any measure. They also must agree, if they got no cake, that this value would be zero; i.e., $\mu(\emptyset) = 0$. We also assume that all of the measures are defined on any piece of cake which is cut.

- Pieces can be repeatedly cut without diminishing total value. This allows us to ask Tom to cut the cake without losing value in the process. More generally, if A has been cut into two pieces B and C, so that $A = B \cup C$ with $B \cap C = \emptyset$, then $\mu(A) = \mu(B) + \mu(C)$ where μ is any player's measure. This condition is called *finite additivity* for the measure μ. For almost all of the algorithms we will encounter, finite additivity will suffice. Nevertheless, there will be algorithms which require the measures to be *countably additive*. This assumption would allow us to cut the cake in an infinite sequence of cuts without losing value. This requirement will be obvious when it appears.

- If a piece A has positive value a to a player, and if $a_1 > 0$ and $a_2 > 0$ with $a_1 + a_2 = a$, then the player can cut A into two subsets which have values a_1 and a_2 respectively.

This agreement allows us to give instructions such as: "Cut the cake into two pieces you consider to have value 1/3 and 2/3 respectively, and then cut a piece of value 1/2 from the larger." We will not question how the player may decide to make such cuts, but simply assume it can be done.

Summarizing our agreements about our measures, if μ is the measure of any player, then we demand that the following conditions hold:

(i) $\mu(\emptyset) = 0$ and $\mu(X) = 1$.

(ii) If $A = B \cup C$ and $B \cap C = \emptyset$, then $\mu(A) = \mu(B) + \mu(C)$.

Commonly used measures, called probability measures, have these properties. For this reason, probability measures (as opposed to other kinds of measures not having these properties) are associated with the cake-cutting problem.

With this notation we can now more formally state the critical features of finite algorithms, which are the ones Steinhaus had in mind when he asked about the number of cuts required for simple fair division. A finite algorithm is understood to have the following properties:

- At each step, any player, say P_i, may be asked to cut an existing piece A into two pieces $A = A_1 \cup A_2$ of specified non-negative sizes a_1 and a_2 such that $a_1 + a_2 = \mu_i(A)$. If the sizes are specified by the algorithm, they are provided so that, if the directions are followed by all cutters at all stages, a fair share will result for each player. If a cutter chooses not to cut the specified values (and nobody other than the cutter can know that), it may be at the risk of his or her fair share. But the fair share of no other player is put at risk.

- The cut is made without consulting any of the non-cutters about the sizes of the resulting pieces A_1 and A_2. Instructions such as "P_1, you cut a piece both you and P_2 consider more than 1/2" are not allowed. Generally we will not even know how such a piece might be found. The cut is made on the basis of only the cutter's judgment.

- Because there is no consultation, it may be the case that for any non-cutter who views piece A as having value $\mu(A)$, and for any two non-negative numbers b_1 and b_2 such that $b_1 + b_2 = \mu(A)$, $\mu(A_1) = b_1$ and $\mu(A_2) = b_2$. In other words, any possible values for A_1 and A_2 could be placed on these pieces by the non-cutters, as long as the values are consistent with their assessment of piece A. (This assumption will be used repeatedly.)

- After a cut is made, all players will give their evaluations of the two new pieces, and on the basis of this information the algorithm specifies the next step. If any player chooses not to give a truthful answer, it does not affect any other player's guarantee of a fair share.

- The above procedures are repeated a finite number of times to accomplish fair division.

In particular, the Moving Knife Algorithm is **not** a finite algorithm, because each player must decide at each moment whether or not the piece to the left of the knife is as large as $1/n$. An infinite (in fact uncountable) number of evaluations is required by each player.

The requirement that all players give exact evaluations of the two new pieces after each cut is sufficient to determine the next step in any algorithm we will encounter. There will be times when not all of that information will be used. For example, it will suffice at times to know whether or not a player thinks a piece is worth (for example) 1/3 without requiring an exact evaluation. It will be apparent when such instances occur.

In counting the number of cuts required by a finite algorithm we need to agree on what gets counted. We will count cuts, not decisions, since our main interest is how many pieces will be required for fair division. Thus, the evaluations the non-cutters give for new pieces after a cut are decisions, not cuts, and will not be counted.

Another important agreement is required, illustrated by the Trimming Algorithm. In order to dispatch the first participant with a fair share, a piece is cut and every other player but one is given the opportunity to trim the piece as it is passed through the players. The only cut that is used for the division is the last one, and in fact the other cuts only serve to generate more unnecessary pieces at the next stage.

So, why not have the players mark the cake by indicating how they would trim the piece if that is required and only cut at the last mark? It is clear that not to count those marks as cuts would be contrary to Steinhaus' view. Indeed, if the marks are not counted, then the Trimming Algorithm is complete with $n-1$ cuts. It is impossible to do better, and there are no "interesting questions" that Steinhaus suggests we explore for finite algorithms. Marking the cake is another physical operation, just like a cut. All such marks will be tallied as cuts.

Finally, we will not question how a player decides to make a cut. We will simply assume that directions given such as "cut a third" or "diminish one piece so it is the size of another" can be followed with the cutter making one cut.

2.2 One Cut Suffices

When counting the number of cuts required by an algorithm, the One Cut Suffices Principle is used repeatedly. Let us illustrate it by returning to the Trimming Algorithm for three persons. What is the largest number of pieces of cake that could result? The worst case scenario occurs when Tom has cut the first piece, Dick has trimmed it, and someone (either Dick or Harry), let's assume Dick, has claimed the trimmed piece. At that point Tom and Harry must divide two unclaimed pieces in halves between them.

If two persons are to divide equally a cake which is already in two pieces and they wish to do it using "Cut and Choose," must they divide each piece separately? The answer is "no," one cut will suffice. Let us say Tom is cutting, and to have some numbers to talk about, let us assume he thinks one piece is worth 2/3 (which is the case in the Trimming Algorithm) and the other (Dick's trimming) is worth 1/12. Then he should get a piece he thinks is $1/2(2/3 + 1/12) = 9/24$ of the cake. With just one cut, he can cut the larger piece into pieces he values at $9/24$ and $2/3 - 9/24 = 7/24$. He can now offer Harry his choice of the piece he values at 9/24 or the two combined pieces he values at 7/24 and 1/12 (note $7/24+1/12 = 9/24$). Tom doesn't care which choice Harry makes, so Cut and Choose has been accomplished with one cut, not two. Note, then, that the Trimming Algorithm for three persons requires at most three cuts, and in some cases will require all three.

Here then is an important tool when considering how many cuts are needed for fair division.

One Cut Suffices

If a person judges n pieces of cake to have values $a_1, a_2, \cdots, a_n$ respectively and is asked to cut the overall holding of the n pieces in the ratio of $b_1 : b_2$, where $b_1 + b_2 = a_1 + \cdots + a_n$, then one cut suffices to accomplish division in that ratio.

Suppose I think four pieces have value .25, .10, .20, and .15 for a total value of .70, and I am asked to create holdings which I judge to have values .30 and .40. Then I can cut the piece worth .10 in half with one cut and have holdings worth $.25 + .05 = .30$ and $.05 + .20 + .15 = .40$ as required.

More generally, having chosen b_1 and b_2 so that $b_1 + b_2 = a_1 + a_2 + \cdots + a_n$, find a_{j+1} so that $a_1 + a_2 + \cdots + a_j \leq b_1 < a_1 + \cdots + a_j + a_{j+1}$ and cut the piece worth a_{j+1} in portions of size $b_1 - (a_1 + \cdots + a_j)$ and $a_{j+1} - (b_1 - (a_1 + \cdots + a_j))$. The first of these two pieces can be combined with the j pieces with values $a_1, \cdots, a_j$ to create a share with value b_1. The remaining pieces have total value b_2.

2.3 On to the Count

Now let us do the actual counting for some of the algorithms Steinhaus had in mind.

Number of Cuts Required in the Trimming Algorithm

Let us start by counting how many cuts might be required to divide a cake fairly among n players using the Trimming Algorithm. Recall that players are eliminated one at a time after all but one have had an opportunity to cut the cake. Thus, to reduce the problem from n to $n-1$ players, it may require as many as $n-1$ cuts if all players exercise their option to trim the piece being passed down the line of players. So, with $n-1$ cuts, one player is sent away happy and the process is repeated for the remaining $n-1$ players. Reasoning as before, it could take as many as $n-2$ cuts to eliminate the next player. This is an application of the One Cut Suffices Principle. Although on the second round of cuts the $n-1$ persons may start with $n-1$ pieces of cake produced by the trimming in round one of the algorithm, not all pieces need to be cut and divided separately. Each division required by the cutters can be accomplished with one cut. Continuing until the last two persons perform Cut and Choose, we see that the number of cuts required can be as many as $(n-1) + (n-2) + \cdots + 1 = (n^2 - n)/2$. For 100 players, 4950 cuts may be required. You probably won't see the method used at weddings or other large gatherings, but the mathematical question is still worth considering as Steinhaus suggested.

Number of Cuts Used in the Successive Pairs Algorithm

An analysis of the number of cuts used by the Successive Pairs Algorithm before using the One Cut Suffices Principle gives:

number of players	number of cuts used
2	1
3	5
4	23
$\vdots$	$\vdots$
n	$n! - 1$

This follows from the fact that when the nth player appears, each of the $(n-1)$ previous players holds $(n-2)!$ pieces and each of these pieces is cut into n parts. Thus $n!$ pieces result from $n! - 1$ cuts.

A comparison of the number of cuts required to fairly divide the cake among n players by the two algorithms is now possible.

n	2	3	4	5	6	$\cdots$
Trimming Algorithm	1	3	6	10	15	$\cdots$
Successive Pairs Algorithm	1	5	23	119	719	$\cdots$

We see that the Trimming Algorithm is much more efficient because $n^2/2$ grows much more slowly than $n!$ and these two terms dominate the respective counts we performed.

But, with a little alteration, we can make the Successive Pairs Algorithm much more efficient by applying the One Cut Suffices Principle. To divide the cake among three players, each person winds up with two pieces. When the fourth player shows up, we previously had each of the three players divide each of their two pieces in four equal parts, requiring a grand total of 18 cuts, each player making six new cuts. But all we wanted the three cutters to do was to create four equal portions. Using the One Cut Suffices Principle, that can be done with three, not six, cuts. So for four persons $1 + 4 + 9 = 14$ (not 23) cuts suffice.

For $n = 5$, each of the four must present their holdings in what are considered equal fifths — a task that each can do with four cuts. We now use $1 + 4 + 9 + 16 = 30$ (not 119) cuts for five persons. Continuing, we see that for n players $1 + 2^2 + 3^2 + \cdots + (n-1)^2 = (1/6)n(n+1)(2n+1)$ cuts suffice (not $n! - 1$). We have made a big improvement because $n!$ grows much more quickly than $n^3/3$, which is the dominant term in our revised count. A revised Successive Pairs line in the table above would be $1, 5, 14, 30, 55, \cdots$ — much better than the unrevised line, but still not as good as the Trimming Algorithm numbers (because n^3 outgrows n^2). This raises the issue of trying to improve on the $n^2/2$ growth resulting from the Trimming Algorithm. Such improvement is possible using an entirely new algorithm.

2.4 Using Fewer Cuts with the Divide and Conquer Algorithm

In the Trimming Algorithm for n players, the first task is to send one player away satisfied with the piece received while the others are willing to play for their fair share of what is left. After at most $n - 1$ cuts, the players fall into two categories, the one sent away happy and all the rest who still have to divide the remainder. In other words, the n players are split into two groups of size 1 and $n - 1$. Could they be split somehow into other size groups where one group would be happy taking a share from one portion of the cake while the other group would be happy with a share of the complementary portion of cake? In such situations it is often most efficient to make the groups as nearly equal in size as possible.

This is what the Divide and Conquer Algorithm will do, and we will see an improvement in the number of cuts required for simple fair division. This algorithm was first described by Even and Paz [EP]. We will use the fact that to divide a cake for one person zero cuts are required (that person gets it all), for two persons one cut suffices (Cut and Choose), and for three persons three cuts suffice (Trimming Algorithm). Let us examine how we can use that information to divide the cake among four persons. Suppose we ask three of the four to make parallel cuts so that the portion to the left of their cut is 1/2 of the cake by their estimations.

We are now looking at three parallel cuts as shown in Figure 2.1. Ask the person who did not cut which of the two portions defined by the middle cut he or she thinks is larger. Suppose the answer is "I prefer the piece to the left side of the middle cut." Then the non-cutter and the player making the left-most cut can divide the piece of cake to the left of the middle cut, while the other two divide the part to the right of that cut.

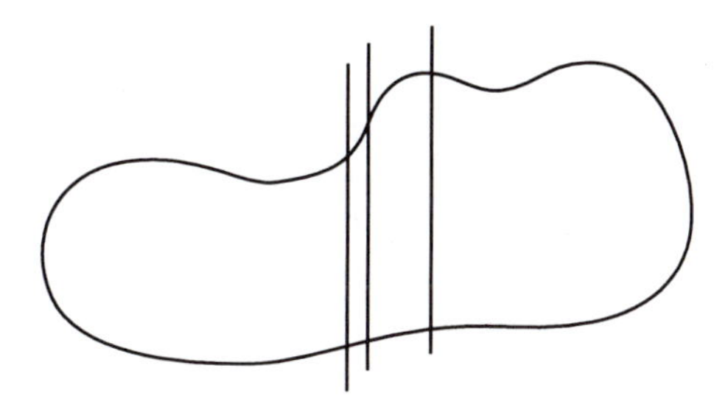

Figure 2.1.

Convince yourself that all four players will now get at least a half (using Cut and Choose) of a piece they think is at least half of the cake; i.e., a piece worth at least 1/4. It is important to note that the middle cutter is indifferent about the two pieces and is willing to be in either of the two groups. (Decide what should be done if the non-cutter prefers the right piece rather than the left or if the cuts of some of the players agree.)

Now, with three cuts, we have reduced the division among four persons to two sub-problems, each of which is to divide cake between two persons, a task that can be done by Cut and Choose. What is our total cut count? Three cuts (for the parallel cuts) plus one cut each for the two-person divisions are required, i.e., $3+1+1=5$. This count beats the Trimming Algorithm by one.

For five persons, we want to divide the players in groups as nearly equal in size as possible. That means we want to reduce the problem to subgroups of size 2 and 3. Ask four players to make parallel cuts dividing the cake into the ratio 2 : 3. Now ask the non-cutter if he or she thinks the portion to the left of the *second* cut is more or less than 40% of the cake. If the answer is "more," let the non-cutter and the player making the left-most cut share that piece. If the answer is "less," let the two players making the two left-most cuts share the piece to the left of the second cut. In either case the remaining three share the piece to the right of the second cut. All five are willing to agree to that. Again note that the person who made the second cut is willing to go with either group.

So, with four cuts, our problem is reduced to a two-person division and a three-person division. Thus, the five-person division is accomplished with four cuts (for the parallel cuts) plus one cut (for the two-person problem) plus three cuts (for the three-person problem) for a total of eight cuts. This has improved on the ten cuts required to divide the cake among five persons using the Trimming Algorithm.

With these examples in mind, we can now state the general procedure for the Divide and Conquer Algorithm. If there are $n = 2k$ players, have all but one divide the cake in the ratio $k : k$ with parallel cuts. Divide the cake into two portions, piece A which lies to the left of the middle cut, and piece $X - A$. The non-cutter chooses X or $X - A$ and shares that portion with the $k-1$ players whose cuts fall in the chosen piece. The other k players share the complementary portion. If there are $n = 2k+1$ players, have all but one divide the cake in the ratio $k : (k+1)$ with parallel cuts. Ask the non-cutter if he or she prefers to share the piece to the left of the kth cut as $k/(2k+1)$ of the cake or to share the piece to the right of the kth cut as $(k+1)/(2k+1)$ of the cake. Depending on the answer, the players are divided into two groups of sizes k and $k+1$. By the inductive process we are using, we will know how to deal with the two new smaller problems since they have already been solved. The following table illustrates the method for small numbers of players:

Number of Players	Method	Number of Cuts
$n = 1$	No division is required.	0
$n = 2$	Use Cut and Choose.	1
$n = 3$	Use 2 cuts to reduce to $2 - 1$ cases.	3
$n = 4$	Use 3 cuts to reduce to $2 - 2$ cases.	$3 + 1 + 1 = 5$
$n = 5$	Use 4 cuts to reduce to $2 - 3$ cases.	$4 + 1 + 3 = 8$
$n = 6$	Use 5 cuts to reduce to $3 - 3$ cases.	$5 + 3 + 3 = 11$
$n = 7$	Use 6 cuts to reduce to $3 - 4$ cases.	$6 + 3 + 5 = 14$
$n = 8$	Use 7 cuts to reduce to $4 - 4$ cases.	$7 + 5 + 5 = 17$
$n = 9$	Use 8 cuts to reduce to $4 - 5$ cases.	$8 + 5 + 8 = 21$
$n = 10$	Use 9 cuts to reduce to $5 - 5$ cases.	$9 + 8 + 8 = 25$

If we denote by $D(n)$ the number of cuts used for the Divide and Conquer Algorithm, we can see that $D(2k) = 2k - 1 + 2D(k)$ and $D(2k+1) = 2k + D(k) + D(k+1)$. These equations are both given by the single equation $D(n) = n - 1 + D(\lfloor n/2 \rfloor) + D(\lceil n/2 \rceil)$. As usual, $\lfloor\ \rfloor$ and $\lceil\ \rceil$ are the floor and ceiling functions respectively. We also have $D(1) = 0, D(2) = 1$ and $D(3) = 3$.

Using induction an exact formula for $D(n)$ is given in Exercise 2.6, which shows $D(n)$ is roughly $n \log_2 n$. The Divide and Conquer Algorithm for large n gives a significant reduction in the number of cuts required over all previously considered algorithms. We finally formally state this new algorithm:

Divide and Conquer Algorithm [EP], [RW1]

If $n = 1$, the player receives all the cake.

If $n \geq 2$, and:

(a) If $n = 2k$, have all but one player divide the cake X by parallel cuts in the ratio $k : k$. The non-cutter chooses either piece A to the left of the middle cut as at least half of cake X or piece $X - A$ as at least half. The non-cutter shares the chosen piece with the $k - 1$ players whose cuts fall within the chosen piece using the algorithm for $n = k$. The remaining k players will perform the algorithm for $n = k$ on the complementary portion. (Ties are resolved arbitrarily.)

(b) If $n = 2k + 1$, have all but one player divide the cake by parallel cuts in the ratio $k : k + 1$. The non-cutter chooses either piece A to the left of the kth cut as at least $k/(2k + 1)$ of X or piece $X - A$ as at least $(k + 1)/(2k + 1)$ of X. In the first case the non-cutter shares A with the $k - 1$ players whose cuts fall within A using the algorithm for $n = k$. In the second case the non-cutter shares $X - A$ with the k players whose cuts fall within $X - A$ using the algorithm for $n = k+1$. In both cases the other players share the complementary piece using the algorithm for $n = k + 1$ or $n = k$ respectively.

By the nature of the recurrence any improvement in the number of cuts for small values of n will automatically produce improvements for larger n. We will see such improvements for $n = 4, 5$, and 6.

In an exercise we explore what might happen if we use cuts other than near-halves. For example, we might always ask the n players to cut in the ratio $3 : n - 3$ and separate our players into groups of 3 and $n - 3$. We will see in Exercise 2.2 that this is not as efficient as cutting near-halves. To "divide and conquer" don't leave any new subgroup any larger than necessary.

2.5 Two Cuts Are Never Enough

Although Moving Knife gives three persons a fair share with two cuts, the fewest number of cuts used by any of our finite algorithms is three. We will now see that three is in fact the best we can do. No finite algorithm can accomplish fair division for three persons with only two cuts. We will first show that we can always guarantee each of the three persons 1/4 of the cake with two cuts. Then we will show why we can't do better.

An Algorithm Using Two Cuts to Give Each of Three Players at Least 1/4 of the Cake

First have P_1 cut $X = X_1 \cup X_2$ with $\mu_1(X_1) = 1/3$ and $\mu_1(X_2) = 2/3$.

Case 1: If $\mu_2(X_2) \geq 1/2$ and $\mu_3(X_1) \geq 1/4$ have P_2 and P_1 share X_2 and give X_1 to P_3. The symmetric case $\mu_3(X_2) \geq 1/2$ and $\mu_2(X_1) \geq 1/4$ is handled similarly.

Case 2: If $\mu_2(X_2) \geq 1/2$ while $\mu_3(X_1) < 1/4$ then P_2 and P_3 share X_2 and P_1 gets X_1. Again the symmetric case is similar. (Thus if either P_2 or P_3 thinks X_2 is at least a half, the mission has been accomplished.)

Case 3: If $\mu_2(X_2) < 1/2$ and $\mu_3(X_2) < 1/2$ then P_2 and P_3 share X_1 and P_1 gets X_2. All possibilities have been covered and each player gets at least 1/4 using only two cuts.

More Than 1/4 Cannot Be Guaranteed to Three Players with Two Cuts

We now see why 1/4 is the most we can guarantee all three if we are allowed only two cuts. Someone, say P_1, will have to make a first cut $X = X_1 \cup X_2$ and the two non-cutters have no control on the values of X_1 and X_2. We can assume that $\mu_1(X_1) \geq 1/2$. It then may be the case that $\mu_2(X_2) = \mu_3(X_2) = 1/2$ and we have only one cut left.

Case 1: If P_1 cuts either X_1 or X_2 in two portions, it may be the case that both P_2 and P_3 view the two new pieces as exact quarters since they thought both X_1 and X_2 were exact halves. Either P_2 or P_3 will get one of those pieces, so somebody will get no more than 1/4, which is what we are trying to show.

Case 2: If a player other than P_1, let's say P_2, cuts X_1 or X_2 into two pieces, then one of them is considered by P_2 to be at most 1/4. It may be, since P_1 and P_3 did not make that cut, that they both agree with P_2 that this piece is at most 1/4. Thus, someone will get a piece he or she considers at most 1/4.

This completes our argument that we can't guarantee more than 1/4 to all three players with just two cuts. Now we know that three cuts and four pieces are the fewest possible to guarantee each of three players a third using a finite algorithm.

2.6 What is the Best We Can Do for Four Players?

We now know that three players require at least three cuts for any finite algorithm for simple fair division. How about four players? The Divide and Conquer Algorithm gives the most efficient *general* result known at this time. In some isolated lower cases, however, *ad hoc* algorithms that seem to depend on the specific numbers being considered (and hence do not lead to general algorithms for all n) have been described which use fewer cuts than the Divide and Conquer Algorithm. In fact, Even and Paz [EP] give an algorithm for four players that uses only four cuts, whereas the Divide and Conquer Algorithm requires five cuts for the same task. This algorithm is clever but somewhat complicated, so the reader may choose to skip the details which are now given.

An essential feature of this algorithm is the observation that a player is satisfied to get a "big" piece or to give up a "small" piece. In particular, suppose the cake X has been cut into four pieces $A \cup B \cup C \cup D$ and some player P believes $\mu(A) \geq \mu(B)$ and $\mu(C) \geq \mu(D)$. Then necessarily $\mu(A \cup C) \geq 1/2$ and P would be willing to share $A \cup C$ with any other player using cut and choose. For convenience we will say that P *prefers* A *to* B if $\mu(A) \geq \mu(B)$, and *strictly prefers* A *to* B if $\mu(A) > \mu(B)$.

Let the four players be P_1, P_2, P_3, and P_4. To begin have P_1 cut $X = Y \cup Z$ so that $\mu_1(Y) = \mu_1(Z) = \frac{1}{2}$.

Case 1: Not all P_2, P_3, P_4 strictly prefer the same piece.

In this case we may assume $\mu_2(Y) \geq \frac{1}{2}$, $\mu_3(Y) \geq \frac{1}{2}$, while $\mu_4(Z) \geq 1/2$. Then P_2 and P_3 can share Y, while P_1 and P_4 share Z. Only three cuts are needed. We see that disagreement makes the task easy.

Case 2: Players P_2, P_3, and P_4 all strictly prefer the same piece.

In this case we may assume $\mu_i(Z) > \frac{1}{2}$ for $i = 2, 3, 4$.

Have P_1 cut $Y = Y_1 \cup Y_2$ so that $\mu_1(Y_1) = \mu_1(Y_2) = 1/4$. Even though the others view Y as worth less than $\frac{1}{2}$, it could happen that one of them, say P_2, thinks $\mu_2(Y_i) \geq 1/4$ for $i = 1$ or 2. This would be lucky for us, since then we can give this Y_i to P_2, the other Y_j to P_1, and let P_3 and P_4 share Z. Again only three cuts are needed.

The alternative is that none of P_2, P_3, P_4 finds either Y_1 or Y_2 acceptable. This case will require some ingenuity. First, note that although neither Y_1 nor Y_2 is acceptable, one must be preferred to the other (allowing ties), and at least two of P_2, P_3, P_4 must prefer the same piece. We may assume that P_2 and P_3 prefer Y_2; we don't really need to know which one P_4 prefers.

Give Y_1 to P_1 and let P_2, P_3, P_4 share $X - Y_1$. At first glance this seems to require five cuts — the two we have already made plus three more on $X - Y_1$. However, $X - Y_1 = Y_2 \cup Z$ and we already know quite a bit about what P_2, P_3, P_4 think of Y_2 and Z. We must use this information carefully.

Have P_2 cut $Z = Z_1 \cup Z_2$ so that $\mu_2(Z_1) = \mu_2(Z_2) > 1/4$. Let's summarize what we know so far in a table. Let + denote an acceptable piece (worth at least 1/4), — denote a piece worth less than 1/4, and $*$ denote a preferred piece (always compare Y_1 to Y_2 and Z_1 to Z_2, since we don't care about comparing Y_i to Z_j).

	Y_1	Y_2	Z_1	Z_2
P_2	$-$	$*-$	$*+$	$*+$
P_3	$-$	$*-$		
P_4	$-$	$-$		

What could the missing entries look like? Remember that among the missing entries a preferred piece must also be acceptable. We may assume that P_3 prefers Z_1. If P_4 finds Z_2 acceptable we have:

		Z_1	Z_2
(a)	P_3	$*+$	
	P_4		$+$

Give Z_2 to P_4 and let P_2 and P_3 share $Y_2 \cup Z_1$. Since P_2 and P_3 share preferred pieces, they must be satisfied. Also, only four cuts have been used.

Finally, if P_4 finds Z_2 unacceptable (so P_4 must prefer Z_1) we have:

		Z_1	Z_2
(b)	P_3	$*+$	
	P_4	$*+$	$-$

Give Z_2 to P_2 and let P_3 and P_4 share $Y_2 \cup Z_1$. Note that P_3 is sharing two preferred pieces and P_4 has given away two unacceptable pieces. So both will get at least 1/4. Again, only four cuts are needed in either (a) or (b).

2.7 Other Information on the Number of Cuts

Some results, found later in Section 9.3, generalize what we learned in this chapter. For $n > 2$, using only $n - 1$ cuts, there is an algorithm that guarantees each player a piece of value at least $1/(2n - 2)$ and this is the best possible using $n - 1$ cuts. Using more cuts allows us to do better; for $n > 3$, with n cuts for n players, we can give each at least $1/(2n - 4)$ and this is best possible, while with $n + 1$ cuts at least $1/(2n - 5)$ can be guaranteed. It is not known whether the last value is the best possible. Arguments for optimality get much more involved as the number of extra pieces increases. It appears that substantial progress will require fresh ideas.

Since the most that can be guaranteed four persons with three cuts is $1/(8 - 2) = 1/6$, we can conclude that the Even and Paz algorithm of Section 2.6, which uses four cuts for four persons, is the best possible. Beyond this, an algorithm requiring six cuts for five players has been given and proved to have the fewest cuts possible. (See Section 9.2.) Also, an algorithm using eight cuts for six players is known, and this improves on other known algorithms [Web2]. But it is not known whether this is the least possible number of cuts.

The problem of determining in general the fewest number of cuts required for fair division seems to be a very hard one. So far, we know that about $n \log_2 n$ cuts are always enough. Yet the Divide and Conquer Algorithm, which gives that bound, has been improved on in certain low cases. The method used in the Divide and Conquer Algorithm, of reducing a given problem to two equal-sized smaller problems, is one which is used in various settings and seems to repeatedly give the best known bounds for many problems. We have lost bets before, but if we were asked to gaze into the crystal ball, we would place our money against finding a substantial improvement on the $n \log_2 n$ bound. Of course, we would expect that, by taking advantage of fortunate properties for special values of n, algorithms could be found for fair division using fewer cuts than the Divide and Conquer Algorithm uses for those n. (These issues will be explored further in Section 9.4.) Nevertheless, arguing that no algorithm can improve on a given result is, in general, a very tricky business.

2.8 EXERCISES

2.1. (a.) How many cuts are required in the worst case to divide cake fairly among three persons using the Kuhn Algorithm described in Exercise 1.8?

(b) For four persons, the matrix might look like:

$$\begin{matrix} 1 & 1 & 1 & 1 \\ 1 & 0 & 0 & 0 \\ 1 & 0 & 0 & 0 \\ 1 & 0 & 0 & 0 \end{matrix}$$

In this case, one of the last three pieces would be given to the cutter and the other three players would have to perform the three person algorithm on the rest of the cake. How many cuts would this require in the worst case?

(c) What would be the count for the number of cuts required for five players for a similar matrix containing a 4×4 block of 0s?

(d) What seems to be the pattern and how does that compare to other counts we made for other algorithms in this section?

2.2. Determine the number of cuts, $D^*(n)$, required for n players when $4 \leq n \leq 8$ if we give instructions to all but one player to use parallel cuts to divide the cake in the ratio $3 : n - 3$. Create a group of three persons who share the cake to the left of the third cut, while the other $n - 3$ will divide its complement. Use values $D^*(1) = 0, D^*(2) = 1$, and $D^*(3) = 3$. To determine $D^*(n)$ for larger n, use values $D^*(m)$ (not $D(m)$) for smaller m. How do the numbers $D^*(n)$ and $D(n)$ compare? This issue is revisited in Section 9.4.

2.3. Convince yourself that all cases have been considered in the four-cut algorithm for four persons. Also, justify that all of the indicated divisions guarantee each player 1/4 of the cake.

2.4. Show it is possible to guarantee each of four players 1/6 of the cake using only three cuts. (Hint: Have P_1 cut halves and consider cases depending on what the others think of that cut.)

2.5. Might it be the case that it could require more cuts for simple fair division for some number of players n than it would for $n+1$ players?

2.6. Show that $D(n) = nk - 2^k + 1$ where $k = \lceil \log_2 n \rceil$. (Hint: You may want to consider the even and odd cases separately.)

THREE

Unequal Shares

3.1 First Things

Up to this point, we have looked only at problems where players are to receive equal shares. Suppose Ann and Beth are to divide leftover cake, but because Beth had a larger piece yesterday they are to divide the remainder of the cake in the ratio 8 : 5. Can we find a way to guarantee Ann 8/13 of the cake and Beth 5/13 by their own assessments? Our first attempts might try to utilize what we know already for equal shares, and in fact an easy solution can be found using any of our algorithms for simple fair division.

A First Solution

Suppose we simply clone Ann until there are 8 of her and clone Beth until there are 5 of her. We can now use any of the algorithms for simple fair division for 13-persons. If cloning proves to be impractical, let Ann represent 8 persons, each using her measure μ_A, and let Beth represent 5 persons, each using her measure μ_B. They then get the portions awarded to the players using their measure by any 13-person equal shares simple fair division algorithm. Note that this method would work for any ratio $p : q$ where p and q are both integers (i.e., if the ratio is rational), but would not apply if the ratio were, say, $\pi : 13$. How does Ann represent π players? It would be a bit strange if $\pi : 13$ were the actual ratio they agreed to! Nevertheless, even though very unlikely in actual cake division, irrational ratios present a legitimate and interesting mathematical problem, and, as we shall see later, it is an essentially different problem from a rational ratio problem. But getting back to Ann and Beth and the ratio 8 : 5, if we used the most efficient algorithm we know to make the fewest number of cuts possible, we would apply the 13-person Divide and Conquer Algorithm. The number of cuts would be $D(13) = 12+D(6)+D(7) = 12+(5+2D(3))+(6+D(3)+D(4)) =$ 37, (or 36 if we use the 4-person algorithm in Section 2.6). Surely we can do better!

Cut Ones Algorithm — A Better Solution

Why not have Ann simply cut the cake into 13 equal portions in the ratio $1 : 1 : \cdots : 1$? Beth chooses 5 of the pieces, Ann takes the other 8. Clearly, by taking the 5 largest pieces Beth gets at least 5/13 of the cake, even if some of the pieces individually might be less than 1/13. That is a clean, simple solution using only 12 cuts, which is a big improvement on the 37 cuts used when we applied the Divide and Conquer Algorithm. There is not much more to be said, other than to ask whether we could do even better yet in terms of number of cuts.

3.2 A Better Solution Leading to Ramsey Partitions

The notion of Ramsey partitions, which we now describe, provides a substantial improvement on the number of cuts required in most cases [MRW]. In the Cut Ones Algorithm, we required a cut to assign a portion of size 1/13. Maybe we can be more efficient by getting rid of bigger portions of the cake with each cut. So suppose we ask Beth to cut the cake in the ratio 8 : 5. We could then ask Ann what she thinks of those two cuts. In particular, do you think the first portion is worth at least 8/13 of the cake? Do you think the second portion is worth at least 5/13 of the cake? She can't say "no" twice (because $\mu_A(X) = 1$). She might say "yes" twice, indicating she exactly agrees with Beth's cut. If we are so lucky, the division is made with one cut and both are satisfied, each getting exactly the share they had coming by their own assessment. More likely, Ann will say "yes" once and "no" once. If she thinks the first piece is worth at least 8/13, she can have it. Beth takes the other piece, and again the division is accomplished with only one cut. But what if Ann thinks the first piece is not worth 8/13 of the cake? Then she thinks the second piece is worth at least 5/13. Give that piece to Ann and let the two complete the division by sharing the first piece.

Since the original ratio was 8 : 5, Ann was to get 8 parts while Beth got 5 parts. But now Ann has been given a piece she considers worth at least 5/13, that is, worth 5 of the 8 parts she has coming. Now they should divide the remainder in the ratio 3 : 5 in Beth's favor. Beth is willing to proceed, because she gave up a portion she considered worth 5 of the 13 parts and Ann's share has been reduced by that amount. If you can't convince yourself right away that Ann is also willing to continue, we will have more to say about this later in Section 3.3. So we are faced with a new problem which is essentially the same as the original: Divide a cake among two persons in unequal portions in a rational ratio. But we have gained, because our numbers are smaller. Forging ahead, ask Ann to cut the unclaimed piece in the ratio 3 : 5. (The person to receive the smaller portion will cut at each stage and the other will keep the chosen piece.) Now Beth has her choice of taking the 5-share piece, in which case Ann takes the 3 and we are done, or Beth can take the 3-share piece to apply toward her 5-share quota. If that happens, she now has 3 shares, while Ann has 5, so the new ratio on the remaining piece becomes 3 : 2 for Ann and Beth respectively. The diagram on page 38 summarizes the entire procedure.

We will use the subscripts A and B to denote pieces chosen by Ann and Beth respectively. Brackets $\{\ \}$ will denote pieces chosen at earlier stages. Thus, the first branch indicates that Beth cuts the whole cake, worth 13 shares, into pieces worth 8 and 5 shares, and Ann must choose one of them. At the last stage either person can cut, since we are essentially playing Cut and Choose.

After all of this, we see that the process is complete with 5 or fewer cuts, much better than our 12 used in the Cut Ones Algorithm. If all five cuts are used, six pieces result with values (5,3,2,1,1,1). This partition of 13 has some very interesting properties, which we will later develop in some detail. It has what has been called a "Ramsey" property, hence the name *Ramsey partition* of 13. In a classical Ramsey theory setting, a mathematical object (e.g., the set of edges in a graph) is divided into two disjoint subsets with the property that if you can't find what you are looking for in one subset, you are sure to find it in the other. In our case, we could ask either of Ann or Beth to cut the cake in the ratios 5 : ③ : 2 : ① : 1 : 1 and ask the non-cutter to circle the pieces she would accept at the advertized value assigned to them by the cutter. Let us say, for example, that Ann cuts and Beth would accept the following circled pieces at advertized value: 5 3 2 1 1 1. Beth is to get 5 parts, but likes only a total of 4 of those cut. But if we can't find circled pieces which sum to 5, we are sure to find pieces that are uncircled which sum to 8 and that is Ann's fair share. In the example we are looking at, Ann can take uncircled pieces of size 5, 2, and 1. She is happy with them, since she cut them at those values! Beth is happy for her to take them, because to her none of them is worth its advertized value. So Beth thinks Ann is taking a total value less than 8 parts. This means that

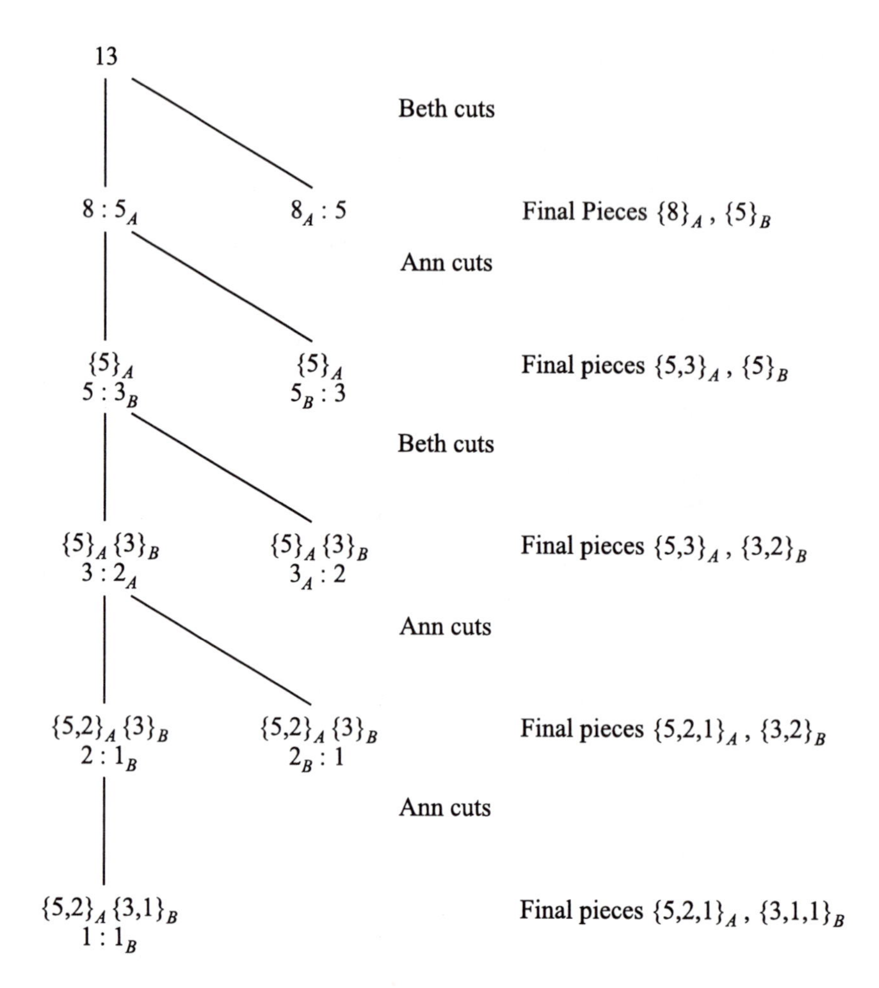
13
Beth cuts
8 : 5_A
8_A : 5
Final Pieces {8}_A , {5}_B
Ann cuts
{5}_A 5 : 3_B
{5}_A 5_B : 3
Final pieces {5,3}_A , {5}_B
Beth cuts
{5}_A {3}_B 3 : 2_A
{5}_A {3}_B 3_A : 2
Final pieces {5,3}_A , {3,2}_B
Ann cuts
{5,2}_A {3}_B 2 : 1_B
{5,2}_A {3}_B 2_B : 1
Final pieces {5,2,1}_A , {3,2}_B
Ann cuts
{5,2}_A {3,1}_B 1 : 1_B
Final pieces {5,2,1}_A , {3,1,1}_B

the rest of the pieces are worth more than 5 parts to Beth, even though some of them are worth less than their advertized value.

Let us formalize the definition of a Ramsey partition:

> If k_1 and k_2 are positive integers, a partition of the integer $k_1 + k_2$ is said to be a *Ramsey partition* in the ratio $k_1 : k_2$ if and only if, for any subset of terms in the partition, if there are not parts which sum to k_1 in that subset, there are parts which sum to k_2 in the complementary subset.

Try the partition (3,2,2,1,1,1,1,1,1). Circle any terms you wish. You will either find circled terms which sum to 8 or uncircled terms which sum to 5. (The roles of 8 and 5 can be reversed: if you can't get a sum of 5 from circled terms, you can always get a sum of 8 from uncircled terms.) So this partition of 13 is also Ramsey for the ratio 8 : 5.

By contrast, the partition (5,5,4,4,1,1,1) of 21 is not Ramsey in the ratio 8 : 13, because if we circle both fives, we can't get 8 by summing circled terms or 13 by summing uncircled ones. There are 792 different partitions of 21, and 46 of them are Ramsey for the ratio 8:13. These are listed in Table 3.1. Any of the 46 could be used for our problem.

Although either can cut, suppose we ask Ann to cut pieces in the ratios of the terms given in any Ramsey partition and have Beth circle the ones she thinks are worth their advertized value. Either Beth will find she has circled pieces so that a sum of 5 can be realized from them (in which case she takes them) or Ann can take pieces she cut that total 8 in value from the uncircled pieces. In either case, as we argued above, both are fairly treated by their own assessment. Let us formally state this algorithm.

> **Ramsey Partition Algorithm for Unequal Shares for Two Players**
>
> **Step 1.** With the ratio $k_1 : k_2$ given, choose any Ramsey partition for the ratio $k_1 : k_2$.
>
> **Step 2.** Ask either player to cut the cake in the ratios given in the Ramsey partition.
>
> **Step 3.** Ask the non-cutter to indicate which pieces are acceptable, i.e., those considered to have value at least as great as what the cutter assigned to them by the cuts made.
>
> **Step 4.** Assign the non-cutter pieces chosen from the acceptable ones whose values sum to the non-cutter's portion, or assign to the cutter pieces chosen from the unacceptable ones whose values sum to the cutter's portion. The other player takes the remaining pieces.

Table 3.1. The 46 Ramsey Partitions for the Pair 8,13.

1.	(8, 5, 3, 2, 1, 1, 1)
2.	(8, 5, 3, 1, 1, 1, 1, 1)
3.	(8, 5, 2, 1, 1, 1, 1, 1, 1)
4.	(8, 5, 1, 1, 1, 1, 1, 1, 1, 1)
5.	(8, 4, 1, 1, 1, 1, 1, 1, 1, 1, 1)
6.	(8, 3, 2, 2, 1, 1, 1, 1, 1, 1)
7.	(8, 3, 2, 1, 1, 1, 1, 1, 1, 1, 1)
8.	(8, 3, 1, 1, 1, 1, 1, 1, 1, 1, 1, 1)
9.	(8, 2, 2, 1, 1, 1, 1, 1, 1, 1, 1, 1)
10.	(8, 2, 1, 1, 1, 1, 1, 1, 1, 1, 1, 1, 1)
11.	(8, 1, 1, 1, 1, 1, 1, 1, 1, 1, 1, 1, 1, 1)
12.	(7, 1, 1, 1, 1, 1, 1, 1, 1, 1, 1, 1, 1, 1, 1)
13.	(6, 2, 2, 2, 1, 1, 1, 1, 1, 1, 1, 1, 1)
14.	(6, 2, 2, 1, 1, 1, 1, 1, 1, 1, 1, 1, 1, 1)
15.	(6, 2, 1, 1, 1, 1, 1, 1, 1, 1, 1, 1, 1, 1, 1)
16.	(6, 1, 1, 1, 1, 1, 1, 1, 1, 1, 1, 1, 1, 1, 1, 1)
17.	(5, 3, 3, 2, 1, 1, 1, 1, 1, 1, 1, 1)
18.	(5, 3, 3, 1, 1, 1, 1, 1, 1, 1, 1, 1, 1)
19.	(5, 3, 2, 1, 1, 1, 1, 1, 1, 1, 1, 1, 1, 1)
20.	(5, 3, 1, 1, 1, 1, 1, 1, 1, 1, 1, 1, 1, 1, 1)
21.	(5, 2, 1, 1, 1, 1, 1, 1, 1, 1, 1, 1, 1, 1, 1, 1)
22.	(5, 1, 1, 1, 1, 1, 1, 1, 1, 1, 1, 1, 1, 1, 1, 1, 1)
23.	(4, 4, 4, 1, 1, 1, 1, 1, 1, 1, 1, 1)
24.	(4, 4, 3, 1, 1, 1, 1, 1, 1, 1, 1, 1, 1)
25.	(4, 4, 2, 2, 1, 1, 1, 1, 1, 1, 1, 1, 1)
26.	(4, 4, 2, 1, 1, 1, 1, 1, 1, 1, 1, 1, 1, 1)
27.	(4, 4, 1, 1, 1, 1, 1, 1, 1, 1, 1, 1, 1, 1, 1)
28.	(4, 3, 1, 1, 1, 1, 1, 1, 1, 1, 1, 1, 1, 1, 1, 1)
29.	(4, 2, 2, 2, 2, 1, 1, 1, 1, 1, 1, 1, 1, 1)
30.	(4, 2, 2, 2, 1, 1, 1, 1, 1, 1, 1, 1, 1, 1, 1)
31.	(4, 2, 2, 1, 1, 1, 1, 1, 1, 1, 1, 1, 1, 1, 1, 1)
32.	(4, 2, 1, 1, 1, 1, 1, 1, 1, 1, 1, 1, 1, 1, 1, 1, 1)
33.	(4, 1, 1, 1, 1, 1, 1, 1, 1, 1, 1, 1, 1, 1, 1, 1, 1, 1)
34.	(3, 3, 2, 2, 1, 1, 1, 1, 1, 1, 1, 1, 1, 1, 1)
35.	(3, 3, 2, 1, 1, 1, 1, 1, 1, 1, 1, 1, 1, 1, 1, 1)
36.	(3, 3, 1, 1, 1, 1, 1, 1, 1, 1, 1, 1, 1, 1, 1, 1, 1)
37.	(3, 2, 2, 1, 1, 1, 1, 1, 1, 1, 1, 1, 1, 1, 1, 1, 1)
38.	(3, 2, 1, 1, 1, 1, 1, 1, 1, 1, 1, 1, 1, 1, 1, 1, 1, 1)
39.	(3, 1, 1, 1, 1, 1, 1, 1, 1, 1, 1, 1, 1, 1, 1, 1, 1, 1, 1)
40.	(2, 2, 2, 2, 2, 2, 1, 1, 1, 1, 1, 1, 1, 1, 1)
41.	(2, 2, 2, 2, 2, 1, 1, 1, 1, 1, 1, 1, 1, 1, 1, 1)
42.	(2, 2, 2, 2, 1, 1, 1, 1, 1, 1, 1, 1, 1, 1, 1, 1, 1)
43.	(2, 2, 2, 1, 1, 1, 1, 1, 1, 1, 1, 1, 1, 1, 1, 1, 1, 1)
44.	(2, 2, 1, 1, 1, 1, 1, 1, 1, 1, 1, 1, 1, 1, 1, 1, 1, 1, 1)
45.	(2, 1, 1, 1, 1, 1, 1, 1, 1, 1, 1, 1, 1, 1, 1, 1, 1, 1, 1, 1)
46.	(1, 1, 1, 1, 1, 1, 1, 1, 1, 1, 1, 1, 1, 1, 1, 1, 1, 1, 1, 1, 1)

A number of questions are raised. Is there an easy way to recognize a Ramsey partition? How are they generated? Are there Ramsey partitions for $k_1 : k_2$ with fewest or most number of terms? How many terms would there be? Answers to all these questions are known and will be completely developed in Section 11.2. Note that while working on a cake-division problem, we have been led to an excursion into elementary number theory.

Here are some facts which we will justify later:

1. There is a unique Ramsey partition with a largest number of terms, namely $k_1 + k_2$ ones. The division based on this partition is just the Cut Ones Algorithm, and $k_1 + k_2 - 1$ cuts are used.

2. There is a unique Ramsey partition with a fewest number of terms. The number of cuts required for this partition is the same as the sum of the quotients in the Euclidean Algorithm for finding the $gcd(k_1, k_2)$. The example worked out in detail above is the minimal Ramsey partition for $8 : 5$. The following display for the ratio $7 : 12$ shows the connection between the Euclidean Algorithm and minimal Ramsey partitions.
   Euclidean Algorithm for $gcd(7, 12)$:

$$\begin{aligned} 12 &= 1 \cdot 7 + 5 \\ 7 &= 1 \cdot 5 + 2 \\ 5 &= 2 \cdot 2 + 1 \\ 2 &= 2 \cdot 1 + 0. \end{aligned}$$

   The minimal Ramsey partition of $7 : 12$ is 7,5,2,2,1,1,1. Note that the numbers in the Ramsey partition are the numbers 7 and all subsequent remainders. The quotient tells us how many times to repeat a given summand, except that there is always an extra 1.

3. An easy test for identifying a Ramsey partition is the following. First, list the terms from largest to smallest. A partition is Ramsey for the ratio $k_1 : k_2$ if and only if when summing terms in the order they appear, leaving out whichever terms you wish, your sums do not skip over either k_1 or k_2. So, for example, (5,3,3,3,2,2,1,1,1) is not a Ramsey partition for $8 : 13$, because 5+3+3=11 and 5+3+3+3=14, so 13 was skipped; or $3 + 3 = 6$ and $3 + 3 + 3 = 9$, so 8 was skipped. Using this and other information known about these special partitions, a computer search can be developed to find all Ramsey partitions for a given ratio.

We will return to fair division using Ramsey partitions in Section 11.2.

3.3 Cut Near-Halves Algorithm for Unequal Shares

Using Ramsey partitions allowed us to use fewer cuts than used in the Cut Ones Algorithm because we got rid of the cake in larger portions. So why not cut near-halves and get rid of as much as possible? To see whether the idea has merit, let us return to the example of the 8 : 5 ratio. There are $8+5 = 13$ units to be assigned. Allow Beth (the player to get the smaller portion) to cut near-halves in the ratio of 7 : 6. Ann can take her choice of pieces at the assigned value. If she takes the piece Beth valued at 7 units, she then has 1 more unit coming, so the other piece must be divided in the reduced ratio 1 : 5. If Ann takes the portion Beth valued at 6 units, Ann then has two more units coming, so they play in the ratio 2 : 5 on the unclaimed piece. Again, a diagram (p. 43) shows all possibilities (using the same notation as in Section 3.2).

Now we see that 4 cuts is the most ever used, so we have improved on the 5 cuts needed to complete the minimal Ramsey Partition Algorithm for 8 : 5.

We can now state the general procedure.

Cut Near-Halves Algorithm

Step 1. With the desired ratio of $k_1 : k_2$, with $k_1 < k_2$ and $(k_1, k_2) = 1$, let the person to receive the smaller portion cut near-halves. If $k_1 + k_2$ is odd, the portion sizes are $(k_1 + k_2 - 1)/2$ and $(k_1 + k_2 + 1)/2$; if $k_1 + k_2$ is even, the portion sizes are both $(k_1 + k_2)/2$. The non-cutter chooses one of the two pieces and the new ratio becomes $(k_2 - (k_1 + k_2 - 1)/2) : k_1$ or $(k_2 - (k_1 + k_2 + 1)/2) : k_1$ in the odd case depending on which piece is chosen, and $(k_2 - (k_1 + k_2)/2) : k_1$ in the even case.

Step 2. If the new ratio reduces to 1 : 1, the two play Cut and Choose on the unclaimed piece. Otherwise, repeat Step 1 for the new ratio.

Some observations should be made. First, notice in the odd case that since $k_1 + k_2 - (k_1 + k_2 - 1)/2 = ((k_1 + k_2)/2) + 1/2$ is an integer smaller than $k_1 + k_2$ (and the two other cases are checked similarly), a $1 : 1$ ratio must eventually result, so the algorithm is finite. Secondly, the algorithm can be seen to guarantee fair shares in the required ratios. We will illustrate the argument by considering shares in the ratio 8 : 5. Beth cuts X into portions A and B in the ratio 7 : 6 and Ann chooses a piece, let's say A. (The other case is argued similarly.) Ann then thinks the value of A is $7 + \epsilon$ for some $\epsilon \geq 0$, and the new ratio becomes 1 : 5. The algorithm calls for Beth to get 5/6 of B, which she considers to have value 6, so if Beth can in fact get at least 5/6 of B, she will get at least (5/6) $\cdot$ 6 = 5 as required. Ann has value $7 + \epsilon$ already claimed in piece A and the algorithm calls for her to get an additional 1/6 of B, which

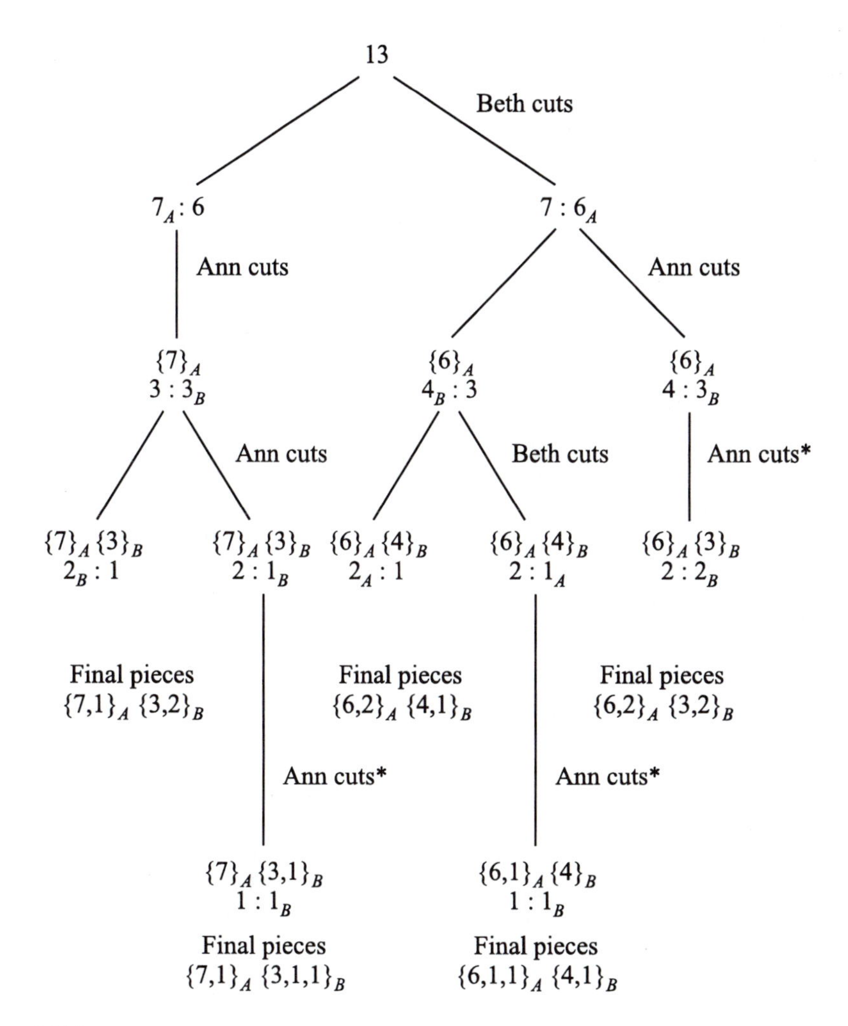

* Either person can cut.

she values at $6 - \epsilon$. So if Ann can get at least 1/6 of B, she will receive at least $(7 + \epsilon) + (1/6)(6 - \epsilon) = 8 + \epsilon(1 - (1/6)) \geq 8$. Note that if $\epsilon > 0$, then Ann will ultimately get strictly more than her fair share. So, if a chooser at any step gets to take a strictly preferred piece (when measured against the cut values), that chooser will ultimately get strictly more than a fair share. This is an example of a recurrent theme, the serendipity of disagreement, which often leads to divisions that players consider more than fair. In Chapter 4 we explore this theme in more detail.

So, will the algorithm give Beth at least 5 portions and Ann at least 8? The answer is "yes," if the algorithm, as it continues, will give Beth at least 5/6 of B and Ann 1/6 of B; i.e., if the algorithm divides B fairly, in the ratio $1 : 5$. We have transferred the question of fairness on X in the ratio $8 : 5$ to the question of fairness on B in the ratio $1 : 5$. This can be repeated, always moving the ultimate issue of fairness on all of X to fairness on a smaller portion of X in a reduced ratio. But finally we reach the ratio of $1 : 1$ and the question becomes, "Does Cut and Choose divide fairly in the ratio $1 : 1$?" Since the answer to that question is "yes," we see that X has been fairly divided in the ratio $8 : 5$.

There is a final observation, which is justified in Section 11.2, that gives an upper bound on the number of cuts used in the Cut Near-Halves Algorithm. Assuming that $gcd(k_1, k_2) = 1$, if $2^{r-1} < k_1 + k_2 \leq 2^r$, then the number of cuts required is no more than r. In the example $8 : 5$ where $2^3 < 8 + 5 \leq 2^4$, we saw that 4 cuts might be required but always sufficed.

Both the Cut Near-Halves Algorithm and the Ramsey Partition Algorithm (in the version where cuts and choices alternate) are special cases of an even more general form. As before, suppose Ann and Beth are to share in the ratio $8 : 5$. We could ask *either one*, say Ann, to cut into pieces $C \cup D$ in the designated ratio $c : d$ where c and d are *any* positive integers with $c + d = 13$. Beth chooses the piece she prefers; i.e., she chooses the first piece C if she thinks it is worth at least $c/13$, but otherwise she chooses piece D thinking it is worth at least $d/13$. She keeps a chosen piece if its designated value is 5 or less, but otherwise she gives the other (unchosen) piece to Ann to keep. We continue with the appropriate new ratio on the unclaimed piece.

As we have seen, the Cut Near-Halves Algorithm would begin cutting in the ratio $7 : 6$, the Ramsey Partition Algorithm would use the ratio $8 : 5$, but we could also try $11 : 2$ or $10 : 3$, etc. Suppose, for example, that Ann cut in the ratio $10 : 3$ and Beth preferred the piece C with designated value 10. Then she would give piece D (with value 3) to Ann and they would now divide C in the ratio $5 : 5$.

The obvious question is: Do we gain anything with this more general algorithm? All of these procedures are recursive in nature, reducing the numbers in the ratio at each step. The worst case is usually when the smaller piece is claimed, since this yields a smaller reduction in the size of the numbers in the new ratio. The advantage of the Cut Near-Halves Algorithm is that it maximizes the size of the smaller piece. Thus we expect the Cut Near-Halves Algorithm to be the most efficient, and it often is. As a matter of fact, it is always at least as efficient as the Ramsey Partition Algorithm, as we will see in Section 11.2. But can we find cases where some other ratio is even better? You might want to try to find such a case for yourself before reading further.

There was a clue in our previous discussion of the $10 : 3$ ratio. If Beth chooses piece D, the next step would be a division in the ratio $5 : 5$, which is the same as $1 : 1$, which trivially requires only one more cut. But what if Beth chooses piece C? She would keep it, and the next step would be to divide D in the ratio $8 : 2$, which is the same as $4 : 1$. Unfortunately, a $4 : 1$ division requires at least 3 cuts (see Exercise 3.6), for a total of 4 cuts, which only ties with Cut Near-Halves. But the possibility that future ratios might reduce is the key to beating the Cut Near-Halves Algorithm.

Let's look at the ratio 7 : 3. It is easily checked that Cut Near-Halves requires up to 4 cuts. But suppose Ann cuts in the ratio 6 : 4. If Beth chooses C, Ann is given D and we continue with the ratio $3 : 3$ on C. Only one more cut is needed here. If Beth chooses D, Ann is given C and we continue with the ratio $1 : 3$ on D. Only two more cuts are needed. In either case at most three cuts are required!

Are there other such examples? Can we beat Cut Near-Halves by more than one cut? Is there an easy way to find the most efficient sequence of ratios? The answers are "Yes," "Yes" (Exercise 3.8), and "That's a very good question."

3.4 More Than Two Players with Unequal Shares

So far we have dealt with only two players. Now we will see that any number of players can be fairly treated with unequal portions, using two-person algorithms repeatedly. For example, suppose Tom, Dick, and Harry are to divide the cake in ratios 8 : 13 : 15, respectively. Have Tom and Dick divide the cake in the ratio 8 : 13 so that Tom initially gets 8/21 and Dick 13/21 of the cake, respectively. Since Harry needs 15/36 of the entire cake, have Harry divide with both Tom and Dick in the ratio 15 : 21, with Harry receiving the smaller portion in each case. Harry now has 15/36 of everything so he is fairly treated. Tom and Dick have $21/36 \cdot 8/21 = 8/36$ and $21/36 \cdot 13/21 = 13/36$, respectively, which were their fair shares. We can now formalize the procedure that shows how to repeatedly use any two-person algorithm to give unequal fair portions to any number of players.

Algorithm for More Than Two Players — Unequal Shares

Suppose $P_1, \cdots, P_n$ are to divide the cake in the ratios $k_1 : \cdots : k_n$ respectively.

Step 1. Assuming a solution for $n-1$ players, partition the cake $X = X_1 \cup \cdots \cup X_{n-1}$ among $P_1, \cdots, P_{n-1}$ in the respective ratios $k_1 : \cdots : k_{n-1}$.

Step 2. Have P_n divide the piece held by each of $P_1, \cdots, P_{n-1}$ with that player in the ratio $k_n : k_1 + \cdots + k_{n-1}$.

Checking for P_i with $1 \leq i \leq n-1$, Step 1 gives player P_i a portion of size $k_i/(k_1 + \cdots + k_{n-1})$, and P_i gets to keep $(k_1 + \cdots + k_{n-1})/(k_1 + \cdots + k_n)$ of that in the later division with P_n in Step 2. Thus, P_i gets $(k_i/(k_1+\cdots+k_{n-1})) \cdot ((k_1+\cdots+k_{n-1})/(k_1+\cdots+k_n)) = k_i/(k_1+\cdots+k_n)$ of the cake as required. On the other hand, P_n gets $k_n/(k_1 + \cdots + k_n)$ of everything as required, and the algorithm accomplishes the division properly. This algorithm is not very efficient in general, and one may find, at least in some cases, *ad hoc* algorithms to do the job with fewer cuts. Furthermore, the number of cuts used can vary considerably based on the order in which the players are listed. (See Exercise 3.4.) In general, regardless of which two-person algorithm is used, the recursion formula for the number of cuts, $N(k_1 : k_2 \cdots : k_n)$, used by the Algorithm for More Than Two Players — Unequal Shares is $N(k_1 : k_2 : \cdots : k_n) = N(k_1 : k_2 \cdots : k_{n-1}) + (n-1)N(k_n : k_1 + \cdots + k_{n-1})$. Nothing approaching a general theory of optimal number of cuts for unequal shares division has been given to date. This problem may prove to be very difficult.

3.5 Unequal Shares in an Irrational Ratio

What if Ann and Beth are to share in the ratio $1 : \pi$, or any irrational ratio? None of our previous methods apply, so new techniques are required. The following can be said:

A. The Near-Exact, Envy-Free Algorithm found in Section 10.3 is a finite unbounded algorithm for division for any number of players in any ratios, rational or irrational. The division procedure is quite involved.

B. There is an extension of Cut and Choose which provides an infinite discrete fair division algorithm for any ratio [RW3].

Suppose the ratio is $a : b$ and assume $a + b = 1$. (Replace $a : b$ by the equal ratio $a/(a+b) : b/(a+b)$ if necessary.) We will use the binary representations of a and b to describe the algorithm, and start with a special case where the binary forms terminate, so that the ratio is rational. Suppose $a = 3/8 = .011$ is the share for Ann, and $b = 5/8 = .101$ is the share for Beth. Have Ann cut $X = A \cup B$ in equal pieces and let Beth choose a piece, say A. Then have Beth make the second cut on the unclaimed piece $B = C \cup D$ in equal parts and let Ann claim either C or D, say C. Let them finally play Cut and Choose on the unclaimed piece D. We claim fair division in the ratio $3 : 5$ (or $3/8 : 5/8$) has been accomplished. In fact, this is simply the Cut Near-Halves Algorithm slightly disguised, and nothing new has been accomplished. For a more interesting example, suppose in binary notation $a = .010010001\cdots$ is Ann's share, and $b = .101101110\cdots$ is Beth's share. Step 1 is to have Ann (the player with a 0 in the first slot) cut the cake in halves and have Beth choose a piece. Beth's remaining share of the unclaimed piece is $b - .1$ or $b - 1/2$, since she has received what she considers at least half of the cake. So the new ratio for Ann to Beth on the unclaimed piece is (.010010001 . . .) : (.001101110 . . .), which is the same ratio as (.10010001 . . .) : (.01101110 . . .). Now the process is simply repeated on the unclaimed piece with Beth cutting halves (because of the leading 0) and Ann choosing. The process continues with a step required for each of the infinite number of slots appearing in the binary representation. Some reflection may convince you that fair division is accomplished in the ratio $a : b$. In any case, this will be fully justified later in Section 11.3. We will also postpone a formal description of the algorithm until that time. The algorithm is not finite, because an infinite sequence of steps is required. Neither is it continuous, because a continuum of decisions is not required as in Moving Knife Algorithms. Algorithms like this are said to be *infinite discrete*.

C. The methods of Section 3.4 apply to irrational as well as rational ratios. So any algorithm accomplishing fair division in irrational ratios for two players, such as that in B, can be extended to provide an algorithm for any number of players in any ratios. A Moving Knife Algorithm for unequal portions will be given in Section 5.4.

If the algorithm used for two persons is infinite, such as in B above, the methods of Section 3.4 provide an algorithm that has one infinite algorithm following another. This is like trying to find your way past any given term in the list $1, 2, 3, \cdots; 1', 2', 3', \cdots$ by moving from left to right. Although 100 steps get me to 100, how do I get to $1'$? Can there be adjustments made in the procedures so that we get only one infinite algorithm and not one followed by another? What we would be looking for is something like replacing $1, 2, 3, \cdots; 1', 2', 3', \cdots$ by

$1, 1', 2, 2', 3, 3', \cdots$. By a rearrangement everything appears in one infinite list, not two. Again, we will say more about this issue in Section 11.3.

3.6 EXERCISES

3.1. What is the minimal Ramsey partition for:
(a) $10 : 23$? (b) $10 : 21$? (c) $1 : 21$?

3.2. How many cuts are needed for a $10 : 23$ division using Cut Near-Halves? How does this compare with the Ramsey partition method in Exercise 3.1 (a)?

3.3. Compare the number of cuts used to fairly divide cake in the ratios $1 : 2 : 7$ (in that order) as outlined in Section 3.4 using the Cut Ones Algorithm and the minimal Ramsey Partition Algorithm.

3.4. Divide the cake in the ratios $2 : 3 : 5$ and $5 : 3 : 2$ in those orders using repeatedly the method of the minimal Ramsey Partition Algorithm for two players as outlined in Section 3.4. Compare the number of cuts used.

3.5. Show that to divide a cake fairly in the ratios $2 : 3 : 4$, as outlined in Section 3.4 using repeatedly the method of the minimal Ramsey Partition Algorithm for two players, 13 cuts are used. Is any other order for the ratios better?

3.6. Show that if the player to receive the smaller portion always cuts, dividing a cake in the ratio $4 : 1$ requires at least 3 cuts no matter what ratios are used.

3.7. Find another example where Cut Near-Halves is not the most efficient method. (Hint: There are no examples with $c + d \leq 8$.)

3.8. Find an example where an alternate ratio is more efficient than Cut Near-Halves by two or more cuts.

FOUR

The Serendipity of Disagreement

4.1 First Things

Most of us prefer to live in harmonious surroundings free of disagreements. If everyone felt exactly as you do on all issues, wouldn't the world be a peaceful, enlightened place! Now, when dividing a cake, a recurrent theme is that when there *is* disagreement, people can be made happier in the end by properly managing the disagreement. This observation is made in the first publication on the subject [Ste1]; to quote Steinhaus, "It may be stated incidentally that if there are two (or more) partners with different estimations, there exists a division giving to everybody more than his due part; this fact disproves the common opinion that differences in estimations make fair division difficult." He attributes this result to Knaster.

The purpose of this section is to give some examples in elementary settings where we see what can be done with disagreement. We have already observed that if Tom and Dick are dividing a cake and think two pieces to be 40%–60% and 70%–30% of the cake respectively, then we can give one a piece he views as 60% and the other a piece he views as 70%. Both have done much better than the 50% they had coming. But that was a fluke of the evaluations, and would presumably not occur under Cut and Choose because the cutter should have cut halves.

There we make the assumption that, to protect himself, Tom (the cutter) cuts pieces he views as equal. These are presented to Dick who takes his choice.

What if there is disagreement? Let's say Dick judges the pieces to be 60%–40%. We will not worry now about what might motivate Dick to share some of his excess, but assuming he will, let's ask him to give 5% away from the 60%, still leaving him with 55%, still more than his fair share by his own estimation.

So let us ask Dick to cut his piece into 12 equal parts. It is sure that if Tom is presented with 12 pieces of Dick's cake (each of which Dick values at 5%), he can't judge all 12 worthless because he views Dick's piece as 1/2. In fact, we see that Tom must view one of the 12 as having value at least $1/2 \div 12 = 1/24$. Why not let Tom have his choice of the 12 pieces, since Dick thinks they are all the same value anyway? Because there was disagreement on Tom's cut, both now have what they consider to be at least $13/24$ of the cake.

4.2 Some Other Examples

Dividing Beachfront Property — Cut Your Own Inheritance

A fair division problem of inherited property illustrates the serendipity of disagreement (see [Ste3]). Suppose n persons have claim to equal shares of a piece of real estate bounded on the north by a lake, on the south by a county road, and on both east and west by property lines perpendicular to the road. In order that all should have access to the road and the lake, they agree that each of the n plots should be connected and also have east and west boundaries perpendicular to the road. Is it possible to find n such plots and assign them so that each is satisfied he or she has received a fair $1/n$ of the property?

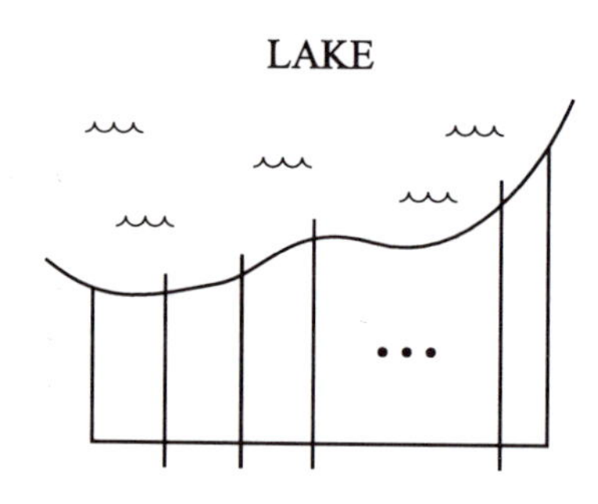

Figure 4.1.

There is a nice, simple procedure that will, in fact, assign to each of the n persons such a plot which he or she "cut." Each of the n persons is given an aerial photograph of the property and told to divide it into n plots of equal value with $n-1$ vertical lines. What we see in Figure 4.1 is what might be produced by one person. Now superimpose all n copies on a master. Figure 4.2 shows how this might look if there were three persons. We see that we can, in this case, let Tom have his first plot, Dick his second plot, and Harry his third plot. Each now has a piece he cut. (Are there other possible assignments that work?) Note there are still two strips of land not used to make these assignments.

Were we just lucky in the order in which these cuts fell or can this always be done? In fact, for any number of persons, and for any order their divisions might appear on the master, there is always a way to assign non-overlapping plots so that each player receives a plot he or she cut. There is a simple method of doing this which is left for Exercise 4.1. In addition, our goal is also to produce an unclaimed strip of land left after everyone has received a self-cut plot if this is possible. This is a bit more complicated.

We will prove the following:

If each of n persons submits divisions of the property as indicated above, and if there are some two persons who have one of their cuts at different places, then there is a way of assigning everyone a plot they cut so that the plots are non-overlapping and the assigned plots do not assign all of the property.

In other words, the only time we cannot have unassigned strips left over is when all n players have placed all $n-1$ of their cuts in complete agreement.

The proof of the italicized statement will be by induction on n. When $n = 2$, the statement is easily seen to be true. Each of the two make only one cut, so to satisfy the hypothesis that cut is not made in agreement (Figure 4.3).

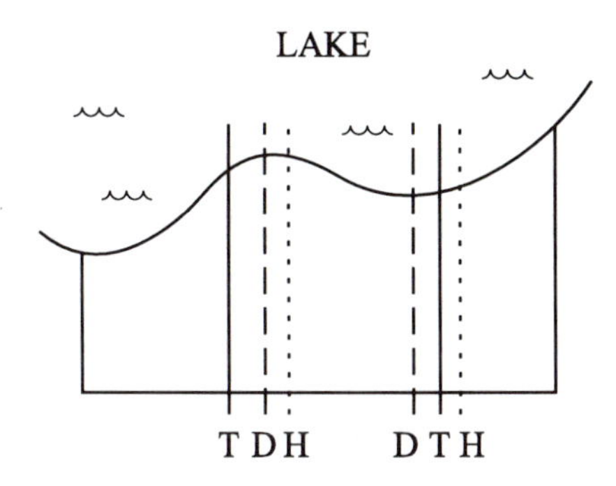

Figure 4.2.

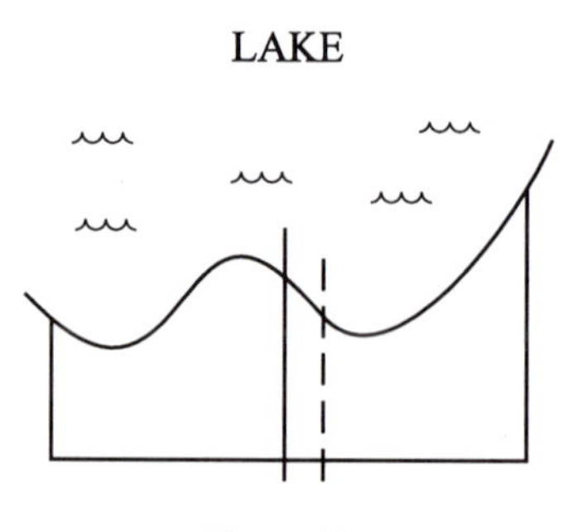

Figure 4.3.

It remains to establish the $(n+1)$st case using the italicized statement above as hypothesis. So, assume all persons $P_1, \cdots, P_{n+1}$, have submitted their cuts, and they are entered on the master. Moving from left to right, proceed to the first line cut by any player. It may be that a number of players, not just one, cut there. Assign that left-most plot to any one player who made that cut, who we label P_1. Remove that plot, and have P_1 drop out. Also remove all of his or her later cuts from the master. Now, temporarily move all other first cuts to the left edge of the remaining property if they aren't there already. (See Figure 4.4 for an illustration when $n = 4$.) In the picture that now remains after the first plot is removed, by the induction hypothesis all can be given non-overlapping plots. But we are still looking for an unassigned strip.

Case 1: There is disagreement in the adjusted plots. Now we need only apply our induction hypothesis to the adjusted figure for the remaining n players.

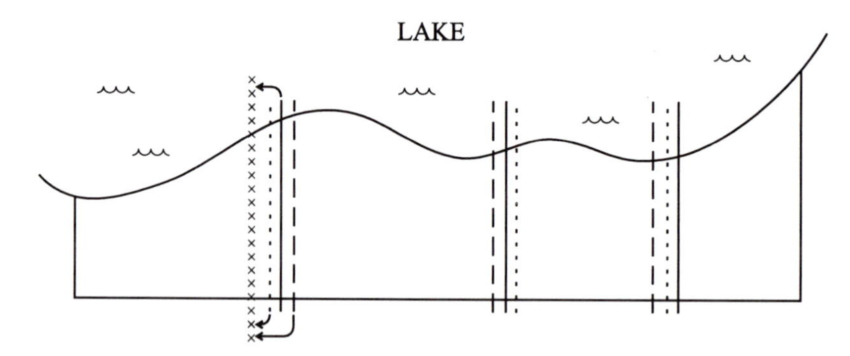

Figure 4.4.

We know that all can be assigned plots defined by their own (temporary) cuts, that they are non-overlapping, and that not all the property is assigned (i.e., there is at least one strip unassigned).

Case 2: There is no disagreement in the adjusted cuts. Then all of the n players still involved made identical second, third, . . . , nth cuts. First suppose some disagreement guaranteed in the original hypothesis for the $n+1$ players occurred on the first cuts. Then, some remaining player P really did get his or her first cut moved leftward under the temporary adjustment. Give the first remaining piece to P. The rest of the plots can be assigned arbitrarily, since all the remaining cuts are identical. At this point everyone has received a plot he or she cut except P, who now has more. In the plot given to P we can now replace its temporary left-most edge by P's original first cut. The strip over which P's first cut was temporarily moved does not need to be used in giving all the players a piece they cut. We now have our unassigned strip.

Otherwise, there was total agreement on the first cut, and the only disagreement in later cuts involved P_1. In this case we could have given the first plot to any player other than P_1, and Case 1 applies.

So what have we observed? Unless all players agree on all of their cuts, they can be given pieces they cut without using up all the land. If the strip has some positive value, it can be sold and the proceeds evenly divided or otherwise disposed of to make a positive contribution to each share. If only two players will disagree on one of their cuts, we can give each player not only a plot he or she cut but an added bonus.

The Trimming Algorithm

Let us return to Steinhaus' statement and see what Knaster had in mind in order to take advantage of disagreement in their Trimming Algorithm. We will for the moment change one step in the algorithm. Rather than give the last person, P_n, the choice of taking or refusing the piece handed him or her, let us ask P_n to reduce the piece, if necessary, to size $1/n$. Thus, P_n will get the same instructions as everyone else. Then, when all the trimming has been done, the piece is given to the last player who reduced its size by trimming. As before, everyone is guaranteed at least $1/n$ of the cake.

Let us assume that because of disagreement some trimming was done, and P_i was the last person to reduce the piece by removing piece E to get piece Y, as shown in Figure 4.5. Then P_i will receive the piece Y and think he or she has an exact $1/n$. Note that all players $P_1, \cdots, P_{i-1}$ think piece Z (which actually might be more than one piece) is at least $(n-1)/n$ in value because of what

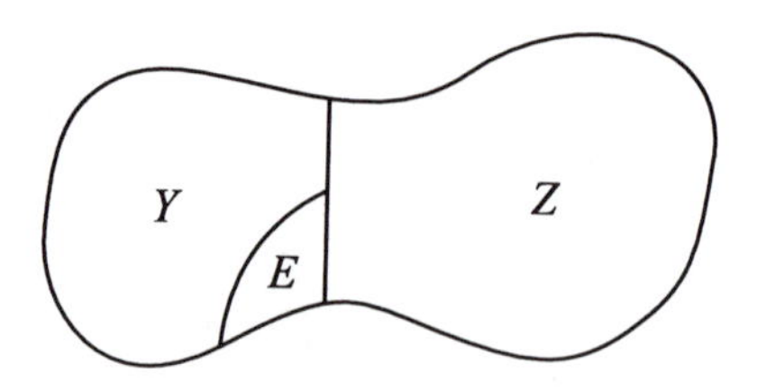

Figure 4.5.

we know about their cuts. If they could get $1/(n-1)$ of Z, that would give them a piece of size $1/n$, which they have coming. On the other hand, players $P_{i+1}, \cdots, P_n$ might all exactly agree with the cut P_i made, so they think $E \cup Z$ is worth $(n-1)/n$ and they need $1/(n-1)$ of $E \cup Z$, not just Z.

How can we proceed from here to give everyone strictly more than $1/n$ of the cake? Suppose we ask P_i to step aside and let the other $n-1$ players divide both E and Z so that each receives at least $1/(n-1)$ of each piece. As we observed above, we could hold back the pieces $P_1, \cdots, P_{i-1}$ get from E, and they would still have what they consider to be $1/n$ of the entire cake. These retained pieces could then be used to produce n smaller pieces, which could be given to the players so that all receive a piece beyond their fair share of size $1/n$.

Some complications exist here. There may be disagreement that running the algorithm would not show. For example, P_1 could cut a portion of size $1/n$ and all others pass it by without trimming. What does that tell us? We know each thinks the piece cut by P_1 has value $1/n$ *or less*, but we don't know which without asking. It could be either, so there may be disagreement that does not produce the slice E we used above to give all more than $1/n$. There is also the issue of some players viewing E as worthless, so getting a part of it does not increase their holdings. These issues will be revisited later in Section 5.2 and Exercise 5.14.

We probably don't need to pursue this further and should now turn to two other examples that show how disagreement can be used. They will be two of the few cases in the book where we divide something that cannot be repeatedly cut without reducing total value.

For the cake-cutting problem we have assumed that the cake can be cut without destroying value, and that values of portions of cake change continuously as a moving knife sweeps over the cake. There are many fair division problems where those assumptions do not apply. For example, suppose two heirs are to inherit a boat. What can be done? Cut and Choose is not a good idea anymore. Since only one can get the boat, it is clear there will have to be something of value (such as cash) other than the boat included in any fair settlement. Our next example shows us what to do.

Dividing an Estate

The following method was also given by Steinhaus [Ste1], when he introduced the subject of cake cutting. Suppose Jill, Abe, and Mary are to receive equal shares of an estate consisting of a car, a boat, and a piano. All have their own opinions about the value of each item. What can be done? Have them each submit independent bids indicating their honest evaluation of each item. An example, along with the division scheme, is indicated in the following table.

	Jill	Abe	Mary
Car	14,300	15,100	13,200
Boat	8,200	7,300	7,300
Piano	3,000	2,500	2,900
Value of the Estate	25,500	24,900	23,400
Fair Share of Estate	8,500	8,300	7,800
Items Received	Boat, Piano	Car	—
Cash Adjustment	-2,700	-6,800	+7,800
Cash Bonus Beyond Fair Share	566	566	566
Lawyer's Fee $2			

The table is largely self-explanatory. They submit their bids (Lines 1–3) and these are totaled (Line 4). Their fair shares will be a third of the values they assign the entire estate (Line 5). Each item goes to the highest bidder (Line 6 — flip a coin in case of ties). If the value of the items received exceeds their fair share, they reimburse the estate the difference. If they don't receive their fair share in items, they are presented the cash difference (Line 7). A negative cash adjustment indicates the person has received items worth more than the fair share value. This amount must be paid back into the estate by the player. At this stage everyone has been given goods and cash worth a third of the estate by his or her own assessment.

In this example, something very fortunate has happened. The amount paid into the estate, $2,700 + $6,800 = $9,500, exceeds the amount paid out, $7,800. So, even after each has received a fair share, there is still enough money left in the estate to give each a $566 bonus (with $2 left over for the lawyer). Where did the money come from? Will the method always guarantee all a fair share? The answer is "*Yes, it will always work and the only time there is no cash bonus produced is when all three agree on the value of each item.*" Any disagreement on any item has the effect of producing cash excess which contributes to the bonus. In a sense, the more they disagree, the happier they will be in the end. Justification of these claims is in the exercises, where we will see exactly how much cash excess the disagreement produces.

Making a Risk-Free Bet

There is a well-known nice variation on Cut and Choose that shows how an informed person like Tom can use disagreement to his advantage. Suppose the Bulls and Knicks are playing, and Tom knows that Dick is a Bulls fan and Harry a Knicks fan. Tom asks Dick, "What do you think will happen?" Dick replies, "I really think the Bulls will win by 5." Let us interpret Dick's honest evaluation

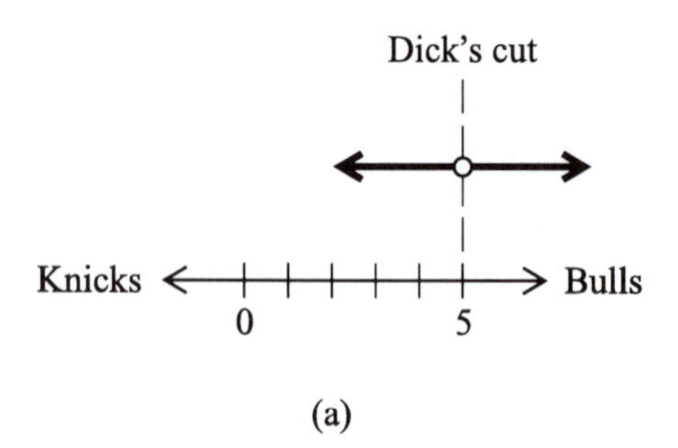

(a)

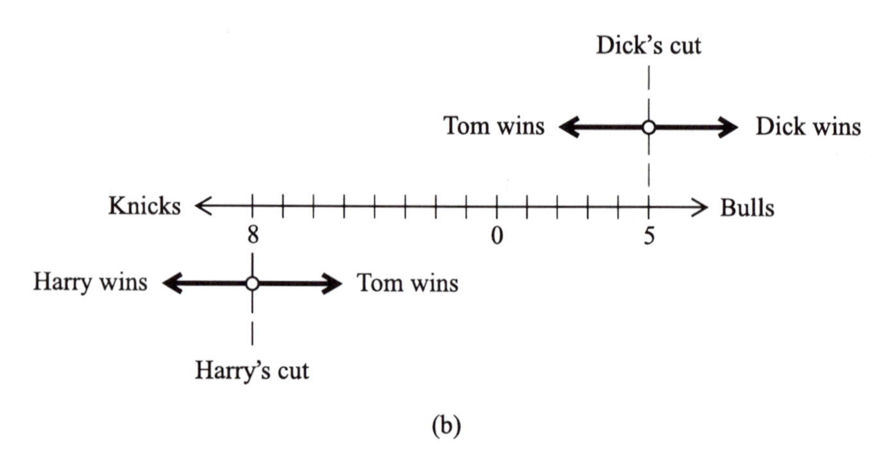

(b)

Figure 4.6.

as the cut shown in Figure 4.6 (a). Dick should be indifferent in taking either bet indicated by the arrows. If the Bulls win by 4 or less, the left bet wins; if they win by 6 or more the right bet wins; if they win by exactly 5, the bet is off.

Next Tom goes to Harry. "Harry, how do you see the game?" Harry replies, "I think the Knicks will win by 8." Now Tom is prepared to use their disagreement.

Suppose Tom takes the two bets as shown in Figure 4.6 (b); i.e., he gives the Bulls to Dick but takes the points Dick offered. Similarly he takes the right-hand bet with Harry. Tom cannot lose under any outcome. If the Knicks win by 9 or more, he wins one bet but loses the other; if the Knicks win by 8, he wins one bet and the other bet is off; if the game ends anywhere between a 7 point Knick win and a 4 point Bull win, Tom wins both bets. And you can easily check the other two cases to see Tom can't lose. He might just as well bet the farm.

In fact, in this case Tom can give both Dick and Harry bets they strictly prefer and still be risk-free. He says to Harry (probably in Dick's absence), "I will give you what you consider better than a fair bet — you can have the Knicks and you win if they win by 8 or more points." What has been done is to move Harry's cut toward Dick's and give Harry a bet he now strictly prefers. His original cut required a Knick win by 9 or more. Repeating the scenario with Dick, he offers a bet Dick wins when the Bulls win by 5 or more, and that is better than 6 or more which Dick has indicated he considers a fair bet. Now Dick and Harry should be eager to make these bets.

With these four simple examples, we have seen how disagreement can be used for the collective good. These are not isolated examples. It is fair to say that differences in evaluations would almost always allow a settlement that gives players shares strictly better than they would get if there were no disagreement. (This issue will be more formally addressed in Chapter 5.) Managing disagreement will be a recurrent theme as we attempt fair division in various settings. A big question is: Can the same principle be found to apply in disputes of more substance? If nations assess situations differently, is there some way to produce harmony rather than conflict with creative management of their differences? With the examples above in mind, it is worth thinking about.

4.3 EXERCISES

4.1. (a) State a method of assignment of pieces which will always ensure that n people dividing beachfront property, as in Example A, will receive a piece of property containing a plot they cut.

(b) Can you produce an example for three players where there is only one way to assign all players a plot they cut?

4.2. (a) Is Tom risk-free on his bets with Dick and Harry if the bets are for different amounts?

(b) Is there a way to make three equal size risk-free bets on the Bulls-Knicks game like those in Example B with three friends simultaneously? four, five, six...? What about unequal size bets?

4.3. (a) In Example C, there was $1700 cash left in the estate after the cash adjustments were made. How much of that $1700 was generated by the car? boat? piano?

(b) Suppose three persons make bids of $a > b > c$ on an item. Write an expression for the cash excess that this item produces in the division method.

(c) From your answer in (b), explain why anytime there is a disagreement on any item, each person can be given strictly more than a fair share of the entire estate.

4.4. Divide the estate in Example C if Jill is to get 50% of the estate, Abe 30%, and Mary 20%. Is more cash excess on a item generated if Jill bids high or if Mary bids high? Why?

4.5. (a) Show that if P_1, P_2, P_3 value items A, B, and C as indicated below, the estate division algorithm in Section 4.2 is not envy-free.

	P_1	P_2	P_3
A	9	8	5
B	8	9	5
C	1	10	14

(b) Modify the algorithm by first adjusting all of the bids, except the highest one, to equal the second highest bid. Now apply the algorithm from Section 4.2. Show that the resulting division is envy-free. (Check against the original, honest bids, not the adjusted ones.)

(c) Show that any such division is envy-free provided all of the bids, except the highest one, are raised to a common level between the values of the highest and second highest bid. (Hint: It suffices to examine the items one at a time.)

4.6. Francis Su and Forest Simmons have suggested the following problem [Su]. Three persons share a three bedroom apartment. The rooms are all different in size, so dividing the rent equally is not fair. All possible ways of assigning values to the rooms can be associated with all points in a triangle with barycentric coordinates (a, b, c), $a \geq 0, b \geq 0, c \geq 0$, where the person occupying room A, room B or room C will pay a, b, or c for their rooms respectively. Normalizing, we may assume $a+b+c=1$. Show that there must be a triple (a, b, c) where each player has a different preferred room at those prices, thus providing an envy-free rent distribution. (Hint: Apply Sperner's Lemma [Ale].)

FIVE

Some Variations on the Theme of "Fair" Division

5.1 Other Interpretations of "Fair"

So far we have attempted, with some success and some failure, to "fairly" divide a cake among n players using different interpretations of "fair." If the problem is to give each player what he or she considers a portion of size at least $1/n$ (i.e., simple fair division), we know a number of ways to do that. If we are to give each an envy-free portion, that is tougher. We might try giving players *strictly* more than $1/n$ of the cake or finding pieces on which there is some agreement of value. Still other interpretations of fair have been formalized and studied. We will now explicitly list some of these (see box on p. 62).

Of conditions (a) – (e), condition (a) is the weakest of all, so that is why it is the easiest to accomplish; if a division satisfies any of the conditions, it will automatically satisfy (a). The implication diagram (Figure 5.1) summarizes the relationships between these different interpretations of fair. An arrow indicates that one type of division implies the other; the absence of an arrow means one type of division can occur without the other being satisfied.

These new, more restrictive interpretations of "fairly" dividing the cake need to be examined separately.

A cake division $X = X_1 \cup \cdots \cup X_n$ among n players where player P_i receives piece X_i is called:

(a) *simple fair division* if $\mu_i(X_i) \geq 1/n$ whenever $1 \leq i \leq n$ (each gets at least $1/n$);

(b) *strong fair division* if $\mu_i(X_i) > 1/n$ whenever $1 \leq i \leq n$ (each gets more than $1/n$);

(c) *envy-free division* if $\mu_i(X_i) \geq \mu_i(X_j)$ whenever $1 \leq i, j \leq n$ (each gets an envy-free portion);

(d) *super envy-free division* if $\mu_i(X_j) < 1/n$ whenever $i \neq j$ and $1 \leq i, j \leq n$ (each feels everyone else receives less than $1/n$), so $\mu_i(X_i) > 1/n$;

(e) *exact* if $\mu_i(X_j) = 1/n$ whenever $1 \leq i, j \leq n$ (each agrees everyone receives exactly $1/n$);

(f) *dirty work fair division* if $\mu_i(X_i) \leq 1/n$ whenever $1 \leq i \leq n$ (each gets a task no more than $1/n$ of the entire job to be done);

(g) *dirty work envy-free* if $\mu_i(X_i) \leq \mu_i(X_j)$ whenever $1 \leq i, j \leq n$ (each gets an envy-free task).

5.2 Strong Fair Division

If all players agree on all cuts made, then there will surely be no way to give each strictly more than $1/n$ of the cake. This will certainly happen if all measures are equal. Thus, strong fair division is not always possible. In Chapter 4, we

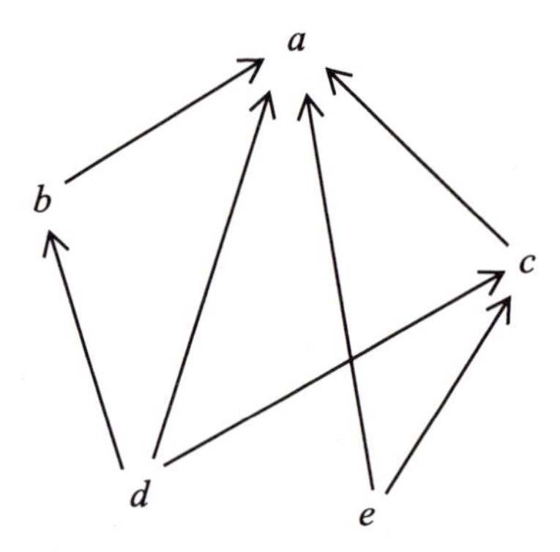

Figure 5.1.

saw that in certain situations when there was disagreement among the players we could give each strictly more than a portion of size $1/n$. It has been proved [DS], [Reb] that if not all of the measures used by n players dividing a cake are the same, then *there exists* a partition which gives strong fair division. So if there are at least two players using different measures, a partition exists that would give all players strictly more than $1/n$ of the cake in their estimations. This assurance of the existence of such sets, however, does not show us how to construct them, which is our main interest.

The problem of getting from the *existence* of the required sets to their actual *construction* hinges on finding at least one set on which at least two players disagree. But knowing only that two players are using different measures (i.e., they don't always agree on the size of a piece of cake), how can we actually find a piece on which there is disagreement? No method for doing this has been given. Think about it — if you know only that two players are going to disagree on some piece of cake, what instructions could you possibly give them to actually carve out such a piece? (See Exercise 5.14.)

Algorithms which actually show how to get pieces that players will think are strictly greater than $1/n$ start with a piece in-hand about which there is disagreement. Where did such a piece come from? We don't know, but if such a piece is available, then strongly fair pieces can be constructed. We will show how the algorithm described by D.R. Woodall [Woo2] produces such pieces.

Woodall Algorithm for Strong Fair Division

The algorithm starts with a piece about which there is disagreement. So let us for the moment assume there are only two players, Tom and Dick, and a piece of cake $A \subseteq X$ for which $\mu_T(A) = a > b = \mu_D(A)$. The strategy will be to find two pieces on whose relative sizes they disagree, give each of them the piece he prefers, and then let them play Cut and Choose on what is left. Find a rational number p/q with $b < p/q < a$. Now have Tom cut A into p equal pieces, $A = A_1 \cup \cdots \cup A_p$. Next, have Dick cut $X - A$ into $q - p$ equal pieces, $X - A = B_1 \cup \cdots \cup B_{q-p}$.

Now we also know that $\mu_T(A_i) > 1/q$ for $i = 1, \cdots, p$ since $\mu_T(A) = a > p/q$, and $\mu_D(B_j) > 1/q$ for $j = 1, \cdots, q - p$ since $\mu_D(X - A) = 1 - b > 1 - p/q = (q - p)/q$. Also, for some $j, \mu_T(B_j) < 1/q$ (otherwise $\mu_T(X) > 1$) and, for some $i, \mu_D(A_i) < 1/q$ (otherwise $\mu_D(X) > 1$). It follows for this i and j that $\mu_T(A_i) > \mu_T(B_j)$ and $\mu_D(B_j) > \mu_D(A_i)$. We now have those two pieces A_i and B_j on whose relative sizes Tom and Dick disagree, and the rest is easy. Let them play Cut and Choose on $X - A_i - B_j$, and also give A_i to Tom and B_j to Dick. Each now has what he considers at least half of $X - A_i - B_j$, plus strictly more than half of $A_i \cup B_j$. Thus, each has strictly more than a half of X.

There is an important point to be made. Note that $q - 1$ cuts were used, but what was q? If $a = 1/100$ while $b = 1/101$, then p/q was chosen between them, which certainly requires $q > 100$. In other words, we don't know q until a and b are known, and q may have to be very large. We finished with $q - 1$ cuts, but we can give no bound for q. Here then is our first example of an *unbounded finite* algorithm. The algorithm will terminate in a finite number of steps, but we cannot give any upper bound at the outset on how many steps will suffice.

If there are three players, we start by giving each of Tom and Dick, who disagree on the size of a piece A, complementary portions X_1 and X_2 that they both consider strictly more than 1/2, as in the paragraph above. We will now let Harry take some cake from each of Tom and Dick so that all wind up with what they consider more than 1/3. To this end, find an integer k so that $[(2k - 1)/(3k - 1)]\mu_T(X_1) > 1/3$. (How we know there is such a k will be discussed shortly. Assume for now that we have such a k.) Then let Tom divide X_1 into $3k - 1$ equal portions and let Harry choose the k pieces considered the largest. Now Tom is left with a portion of size $\mu_T(X_1) - [(k/(3k-1)]\mu_T(X_1) = [(2k - 1)/(3k - 1)]\mu_T(X_1) > 1/3$ as required. Also, Harry has a portion worth at least $[k/(3k - 1)]\mu_H(X_1) > 1/3\mu_H(X_1)$. The same process is repeated for Dick and Harry, giving each a portion considered strictly larger than 1/3.

How do we find the required k? We need $[(2k - 1)/(3k - 1)]\mu_T(X_1) > 1/3$ and we know that $\mu_T(X_1) > 1/2$. So let us write $\mu_T(X_1) = (1 + \epsilon)/2$; i.e., Tom thinks X_1 is worth $1/2 + \epsilon/2$. We don't know how big $\epsilon/2$ is, but we do know it is positive. We now need a k such that $(2k - 1)(1 + \epsilon)/2 > (3k - 1)/3$ or $\epsilon(6k - 3) > 1$. For large enough k, this inequality will be satisfied. How big k will have to be will depend on how small ϵ is. Recall that Tom makes $3k - 2$ cuts, so again we cannot know in advance how many cuts that will be. It depends on how much bigger than 1/2 Tom values X_1. The process is repeated for Dick. This is another *unbounded* but *finite* step in the algorithm.

For four persons, we follow the pattern. Give the first three players portions they all consider strictly greater than 1/3. Now Ann needs to get what she considers more than 1/4 of the portions held by each of the other three (i.e., more than 1/4 of the whole cake) while still leaving them all a portion they consider strictly more than 1/4. So find an integer k such that $[(3k - 1)/(4k - 1)]\mu_T(X_1) > 1/4$. Have Tom cut X_1 into $4k - 1$ equal pieces and let Ann take the k pieces she prefers. This is repeated with Dick and Harry to complete the four-person problem. (See Exercise 5.2 for the general case.)

We have left some open questions. How do we find a piece on which there is disagreement? Also, is there a *bounded finite* algorithm that will give each player strictly more than a portion of size $1/n$; i.e., an algorithm that we know in advance will not require more than some fixed number of steps?

5.3 Envy-Free Division

In Chapter 1, we discussed envy-free division and described the Selfridge Algorithm for three players. A nice moving knife solution to this problem has also been given. In fact, there are four knives moving simultaneously.

The Stromquist Moving Knives Envy-Free Algorithm for Three Persons

Walter Stromquist described a Moving Knives Algorithm which gives envy-free division for three persons. We will quote Stromquist [Str]:

> "A referee moves a sword from left to right over the cake, hypothetically dividing it into a small left piece and a large right piece. Each player holds a knife over what he considers to be the midpoint of the right piece. As the referee moves his sword, the players continually adjust their knives, always keeping them parallel to the sword (see Figure 5.2). When any player shouts "cut," the cake is cut by the sword and by whichever of the players' knives happens to be the middle one of the three.
>
> The player who shouted "cut" receives the left piece. He must be satisfied, because he knew what all three pieces would be when he said the word. Then the player whose knife ended nearest to the sword, if he didn't shout "cut," takes the center piece; and the player whose knife was farthest from the sword, if he didn't shout "cut," takes the right piece. The player whose knife was used to cut

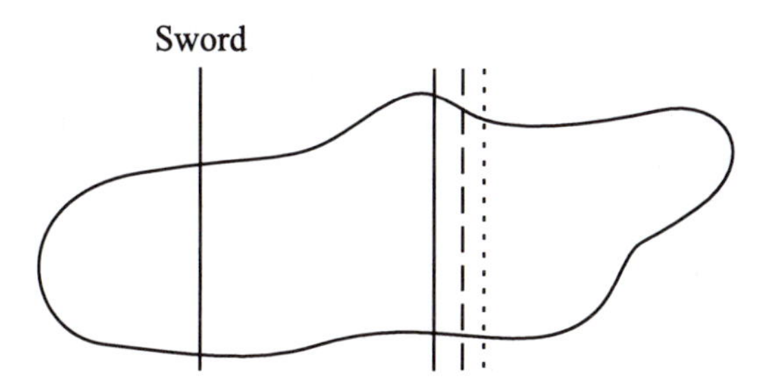

Figure 5.2.

> the cake, if he hasn't already taken the left piece, will be satisfied with whichever piece is left over. If ties must be broken — either because two or three players shout simultaneously or because two or three knives coincide — they may be broken arbitrarily."

You can check that the process generates envy-free assignments (see Exercise 5.3). In his article, Stromquist acknowledges the finite algorithm of Selfridge, Conway, and Guy, and notes that neither method easily generalizes to more than three players. A single moving knife envy-free algorithm is found in Exercise 5.4. We next turn to a recently discovered four-person envy-free solution using moving knives.

The Brams-Taylor-Zwicker Moving Knives Envy-Free Division for Four Persons

The algorithm starts with repeated applications of the following (attributed by Brams and Taylor to A.K. Austin) [BT3].

Moving Knife Algorithm for Cutting a Piece That Each of Two Players Values Exactly One-Half

Have Tom place one knife at the left edge of the cake and another knife parallel to the first so that 1/2 of the cake lies between the knives. He then moves both knives continuously rightward, always keeping half the cake between the knives, until Dick identifies a point where he agrees with Tom.

The question is: Is there sure to be such a point? We might get lucky and find Tom and Dick in agreement on the very first position of the knives. If so, all is well, and we need go no further.

If Dick does not agree that the first position of the knives contains an exact half between them, then let us assume he thinks there is less than half there. As Tom moves the knives, Dick's opinion of the cake between the knives will change continuously even though Tom always considers that portion to be a half. By the time the right-most knife reaches the right edge of the cake, the left knife will be in the position where the right knife started the proceedings.[1] In other words, when the right-most knife reaches the right edge of the cake, the portion between the knives is the complement of the portion between the knives at the

[1] There is a sleeping dog here if there is a section in the center of the cake which Tom considers without value. Then when one knife reaches this section it can continue to move while the other knife is stopped without changing the value between the knives. For this reason, for a position of one knife there may be more than one place to put the second in order to have a half between them. This can be prevented by assuming that as long as two parallel knives over the cake do not coincide, the value of the cake between them is not zero. This assumption is not necessary but provides the fewest complications.

outset. But Dick thought the left piece was worth less than a half, which requires him to value the right piece greater than a half. Since Dick's evaluations changed continuously as the knives moved, and changed from "too small" to "too large," in the spirit of Goldilocks and the Intermediate Value Theorem, it must have been "just right" somewhere along the way. Thus we are sure such a position will always be encountered.

So we have a way to produce two pieces that both Tom and Dick consider exact halves. If they repeat the process on each of those pieces, there will result four pieces which they agree are all exact quarters. These four pieces are the starting point of the next algorithm [BTZ2].

The Brams, Taylor, and Zwicker Moving Knives Envy-Free Algorithm for Four Persons

Step 1. Let P_1 and P_2 cut four pieces $X = X_1 \cup X_2 \cup X_3 \cup X_4$ for which $\mu_1(E_i) = \mu_2(E_i) = 1/4$ for $i = 1, 2, 3, 4$.

Step 2. Ask P_3 to order them in size and assume $\mu_3(X_1) \geq \mu_3(X_2) \geq \mu_3(X_3) \geq \mu_3(X_4)$. Have P_3 cut $X_1 = E \cup X_1^{'}$ so that $\mu_3(X_1^{'}) = \mu_3(X_2)$. ($E$ could be empty.)

Step 3. E is momentarily set aside and P_4 gets first choice of pieces $X_1^{'}, X_2, X_3, X_4$.
Case 1: If P_4 chooses $X_1^{'}$, then the others choose in the order P_3, P_2, P_1.
Case 2: If P_4 does not choose $X_1^{'}$, then $X_1^{'}$ is assigned to P_3, and P_1 and P_2 each take one of the remaining pieces.

Step 4. Whoever holds $X_1^{'}$ is renamed "Non-Cutter" and the other of P_3 or P_4 is renamed "Cutter."

Step 5. Have P_2 and "Cutter" cut $E = E_1 \cup E_2 \cup E_3 \cup E_4$ so that $\mu_2(E_i) = \mu_2(E)/4$ and $\mu_C(E_i) = \mu_C(E)/4$ for $i = 1, 2, 3, 4$.

Step 6. The pieces E_1, E_2, E_3, and E_4 are chosen in the order "Non-Cutter," P_1, "Cutter," and P_2.

The justification that the algorithm produces the required four envy-free portions is easy and similar to that of Selfridge's Algorithm.

Further results have been obtained on envy-free division. An existence proof to show that such envy-free portions always exist for n persons has been given. See [Alo], [DS] and [Woo1]. Subsequently, two algorithms have been given to accomplish such a division for *any* number of players [BT2], [RW4]. Even so, the problem is far from satisfactorily resolved because of the nature of these two algorithms. Both are finite (sure to finish with some finite number of cuts) but unbounded (we can't know in advance any bound on how many cuts may be required). In both algorithms, great use is made of very small pieces and it isn't known in advance how small those pieces will have to be. Both algorithms are quite intricate and are given later in the book. (See Section 10.3.)

This, then, leaves us with an important open question. Can we produce a finite *bounded* algorithm for four or more persons such as Selfridge has provided for three? We are waiting for someone to have a good insight.

Super Envy-Free Division

The notion of super envy-free division has been explored by Barbanel [Bar1]. His result is as follows: Super envy-free division is possible among n players if and only if the measures $\mu_1, \mu_2, \cdots, \mu_n$ are linearly independent; i.e., there are no real numbers $a_1, a_2, \cdots, a_n$ other than $a_1 = a_2 = \cdots = a_n = 0$ for which $a_1\mu_1 + \cdots + a_n\mu_n = 0$. The proof of Barbanel's Theorem is an existence proof and does not give an algorithm for finding the required pieces. However, given pieces on which there is appropriate disagreement, there is an algorithm which produces super envy-free division [Web4]. The proof uses some complicated mathematical machinery that we will take up later in the book. (See Section 10.4.)

5.4 Exact Division

We have already used a Moving Knife Algorithm in Section 5.3 which gives an *exact* division of cake for two players. Have one player move two parallel knives continuously across the cake so that the player always keeps exactly 1/2 of the cake between the knives. At some point, the second player must agree. We will generalize this procedure for values other than equal shares shortly. It will be shown later (see Section 8.3) that there cannot be a *finite* algorithm, *bounded* or *unbounded*, for the same task. However, existence proofs have been given to show that the necessary pieces do exist for *exact* division among n persons [Alo].

Near-Exact Division

While *exact* division with a *finite* algorithm, even among two players, is impossible, we can come arbitrarily close. How might this be done? Suppose we have a large pile of pieces that both players think are small. Start giving out the pieces, and if one player gets too far ahead of the other, sort through the unassigned pieces and locate one that will close the gap.

To be precise, suppose we are required to partition $X = X_1 \cup X_2$ so that $|\mu_i(X_j) - 1/2| < \epsilon$ for $1 \leq i, j \leq 2$, where $\epsilon > 0$ is some prescribed small tolerance which dictates how far we are allowed to miss exact fair division. We can find such pieces, and once ϵ is known, we can do it in a fixed number of steps (which depends only on ϵ).

Have P_1 cut X into pieces that are each considered smaller than $\epsilon/2$, which can be done with $\lfloor 2/\epsilon \rfloor$ cuts. Have P_2 reduce any of those pieces if necessary (again with at most $\lfloor 2/\epsilon \rfloor$ cuts) so that P_2 also considers all pieces smaller than $\epsilon/2$. We now have $X = A_1 \cup \cdots \cup A_N$ for some $N \leq 2\lfloor 2/\epsilon \rfloor$, and $\mu_i(A_j) < \epsilon/2$ for $i = 1, 2$ and $j = 1, 2, \cdots, N$.

Form the vectors $v_1 = (\mu_1(A_1), \mu_2(A_1)) = (x_1, y_1), \cdots, v_N = (\mu_1(A_N), \mu_2(A_N)) = (x_N, y_N)$ and rename them as $w_1, w_2, \cdots, w_N$ according to the following procedure.

1. Set $w_1 = v_1$ and place it with its initial point at (0,0).

2. If the end of w_1, which is at (x_1, y_1), is above the line $y = x$ (i.e., if $y_1 > x_1$), sort through the remaining vectors to find one, say v_i, for which $y_i < x_i$. Such a vector must exist since $1 = \Sigma x_i = \Sigma y_i$. Set $w_2 = v_i$ and place the initial point of w_2 at (x_1, y_1). Relabel A_i, calling it A_2, and vice versa.

3. Continuing this way, any time the last endpoint is above $y = x$, choose an unused vector v_j so that $y_j < x_j$. If the last endpoint is below $y = x$, choose the next v_j so that $x_j < y_j$. Continue until all vectors are placed and a path from (0,0) to (1,1) is formed by the vectors.

Note that the vertical distance from any vector endpoint to $y = x$ is smaller than $\epsilon/2$. Also, since each $x_i < \epsilon/2$, for some k we have $w_1 + w_2 + \cdots + w_k = (a, b)$, where (a, b) lies in the pentagon shown in Figure 5.3 with $|a - 1/2| < \epsilon/4$ and $|b - 1/2| < 3\epsilon/4$. For the required portions, let X_1 and X_2 be the union of the A_i's corresponding to $w_1, \cdots, w_k$ and $w_{k+1}, \cdots, w_N$ respectively.

This process will be extended to n players later in the book using the essential ideas we have just seen. The proof will rely on this intuitive fact (which will be justified for higher dimensions): If you have a large number of short vectors whose sum is the 0 vector, they can be laid end to end in some appropriate order to get a path from the origin back to the origin that never moves very far from the origin. This near-exact division will be used to construct other algorithms, including a finite unbounded algorithm for envy-free division among n players for any n.

In summary, we have noted that sets for *exact* division exist, but they cannot be found by a finite algorithm. Only in the case $n = 2$ do we know how to find them using a *continuous* (Moving Knife) algorithm. However, we can come arbitrarily close to exact division among n players using a finite algorithm.

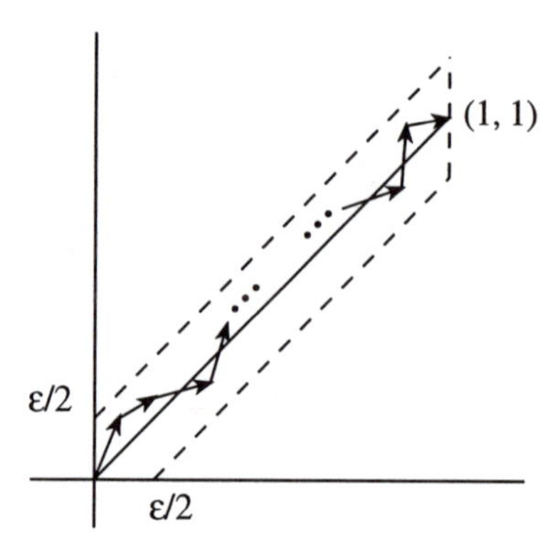

Figure 5.3.

The Near-Exact Algorithm is a handy tool in creating other algorithms. (See, for example, the algorithms in Sections 10.3 and 10.4.) Applications generally have the common feature that there is some disagreement on the value of a given piece, and by dividing various portions in a near-exact manner, shares can be created that more than satisfy the players. Nevertheless, it must be remembered that a drawback to any such procedure is that there will be no absolute bound on the number of cuts such algorithms will require. The following example illustrates the process and anticipates the work to be done in Chapter 10.

Example: Suppose Fiona and Gwen are to divide a cake in the unequal ratio 3 : 2 with Fiona receiving the larger piece. Suppose a cut has been made, $X = A \cup B$, such that $\mu_F(A) = .7$ and $\mu_G(A) = .8$ so that $\mu_F(B) = .3$ while $\mu_G(B) = .2$.

We can take advantage of the disagreement by using ratios different from 3 : 2 on the two pieces, giving Fiona more of B while Gwen gets more of A. If all that we want is strong fair division any simple fair method on the two pieces suffices, but if we eventually want envy-free division where other players are involved, we must use near-exact division.

Getting back to the pieces A and B, a range ratios will work. The fact that both think piece B is less than Fiona's share suggests that we should be able to give all of B to Fiona. Suppose we do this and divide piece A in the ratio 7 : 8. Then Fiona gets $\frac{7}{15} \cdot \frac{7}{10} + \frac{3}{10} = .6266\cdots$ and Gwen gets $\frac{8}{15} \cdot \frac{8}{10} = .4266\cdots$ so each gets the same excess of $.0266\cdots$. Note that it is impossible to simultaneously give both Fiona and Gwen larger portions unless there is further disagreement.

Before leaving the topic of exact division we will give a Moving Knife Algorithm that awards two players pieces about which they agree, even when the shares are unequal.

A Moving Knife Exact Division for Two Players in Unequal Portions

Let us illustrate the algorithm for a specific case. Suppose that Ann and Beth are to divide the cake in the ratio $8:13$. If we ask Ann to move two parallel knives from left to right across the cake, always keeping $8/21$ of the cake between the knives, will it necessarily be the case that somewhere along the way, by the time the right-most knife reaches the right side of the cake, Beth will agree? As we saw above, the answer was "yes" if $1/2$ of the cake was always between the knives. But how about $8/21$?

Unfortunately, the answer is "no"; agreement does not necessarily occur somewhere along the way. (See Exercises 5.5 and 5.6.) But something very fortunate happens. If they never agree that $8/21$ of the cake lies between the knives, then if the process is repeated with Ann keeping $13/21$ of the cake between the knives, at some point Beth must agree. To see why, think of cloning the cake and laying 8 copies of the cake end to end. Place 21 consecutive slices adjacent to each other, as shown in Figure 5.4, each of which has value 8/21 according to Ann's evaluation. Note that since there are 8 cakes, the last slice ends precisely at the right edge of the 8th copy of the cake.

Now we can ask Beth what she thinks of each of the 21 slices. If she agrees with Ann on any slice, we have the desired exact division since the complementary portion of X will be agreed to have value $13/21$. Note that unless the agreement is on the first or last slice, one of the portions on which there is agreement will be composed of two pieces of X. For example, the slice on which they agree (such as the third one shown in Figure 5.4) may actually

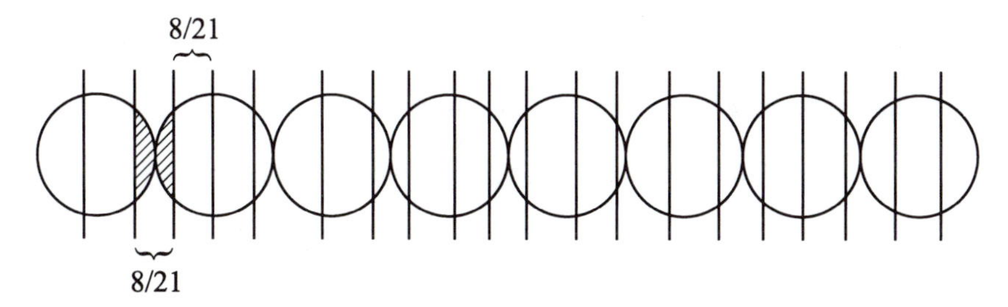

Figure 5.4.

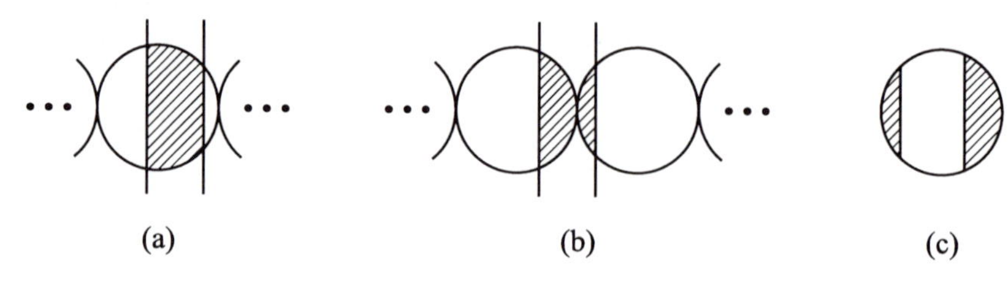

Figure 5.5.

define a portion of X made up of two pieces, one taken from the right edge and the other taken from the left edge of X. In this case the complementary portion is a single piece. On the other hand, if the agreement is on the second slice, the complementary portion is now in two pieces.

But, alas, Beth may not view any of the 21 slices as exactly $8/21$ of X. But she can't view all 21 as worth less than $8/21$, nor can she view all 21 as worth more than $8/21$. (Why?) It follows that there must be two slices S^* and S^{**} that are adjacent, one of which Beth judges to be more and the other one less than $8/21$ of X. Let Ann now place parallel knives over S^* and move them toward S^{**}, always keeping what she considers $8/21$ of the cake between the knives. At some point before the knives reach the position over S^{**}, Beth *must* agree with Ann (because of the Intermediate Value Theorem). The portion which they agree is exactly $8/21$ will look like the shaded region in Figure 5.5(a) or 5.5(b). In the latter case, the shaded portion is equivalent to that in Figure 5.5(c). Note that in all cases one of the two complementary portions on which there is agreement will be a single slice of X. So if there is never an agreement as Ann moves parallel knives across a single copy of the cake always keeping $8/21$ between the knives, there will be agreement somewhere if she keeps $13/21$ between the knives.

This argument easily generalizes to any rational ratio $k_1 : k_2$.

Moving Knife Exact Division Algorithm for Two Players in the Rational Ratio $k_1 : k_2$.

Step 1. Have one player move parallel knives from left to right across the cake, always keeping $k_1/(k_1 + k_2)$ of the cake between them. If at any point the other player agrees, the necessary exact shares are produced.

Step 2. If no agreement is found in Step 1, repeat, keeping $k_2/(k_1 + k_2)$ of the cake between the knives. At some point, there will be agreement and the necessary exact shares are produced.

This method can also be employed for a Moving Knife Exact Division Algorithm for any irrational ratio. We outline the reason why. Assume the cake lies above the interval [0,1] and the irrational ratio desired is $\alpha : 1$, with α irrational. As above, we will show that there must be either a slice of size $\alpha/(1+\alpha)$ or of size $1/(1+\alpha)$ on which they agree. Let (a_n) be a sequence of rational numbers which converge to α. Using the Moving Knife Exact Division Algorithm for Rational Ratio $a_n : 1$, we can establish an interval $[x_n, y_n]$ in [0,1] which provides exact agreement on either a slice of size $a_n/(1+a_n)$ or of size $1/(1+a_n)$. Some subsequence of these intervals have left-hand endpoints which converge to $x_0 \in [0,1]$ and right-hand endpoints which converge to $y_0 \in [0,1]$. So the interval $[x_0, y_0]$ provides exact agreement on a slice of size $\alpha/(1+\alpha)$ or $1/(1+\alpha)$. In either case we have exact division in the ratio $\alpha : 1$.

5.5 Less Is More — The Dirty Work Problem

Bigger is not always better. If we are dividing unpleasant chores, then we prefer a small portion. The dirty work problem is briefly mentioned by Martin Gardner [Gar], but very little has been written about the problem since then.

To be specific, suppose Tom and Dick have to mow a lawn. Can we divide the lawn in two portions and assign the portions to them so that each feels he is doing $1/2$ or less of the work? Will Cut and Choose work? What if Harry offers to help? Then, for example, Moving Knife could be used. Have the knife move from left to right and the first player to say "cut" when he feels exactly $1/3$ of the task is to the *right* of the knife gets that portion to mow. The other two divide the lawn to the left of the cut using Cut and Choose.

We will see in the exercises that the algorithms that we have for simple fair division of the cake can be rather easily altered to guarantee each player a piece he or she considers to be no larger than $1/n$ of the lawn. However, the Selfridge Envy-Free Algorithm for three persons does not quickly lead to an envy-free dirty work solution.

You may wish to review that algorithm before seeing how Reza Oskui has adapted those basic notions to give a very clever method for assigning envy-free tasks to Tom, Dick, and Harry [Osk]. First, envy-free jobs will be assigned on all but trimmed pieces. One player will have an advantage in that assignment, which will not be compromised by the fact that one player is allowed to choose before him on the division of the trimmed pieces. The procedure is more involved than the five-cut Selfridge Algorithm and can require nine cuts.

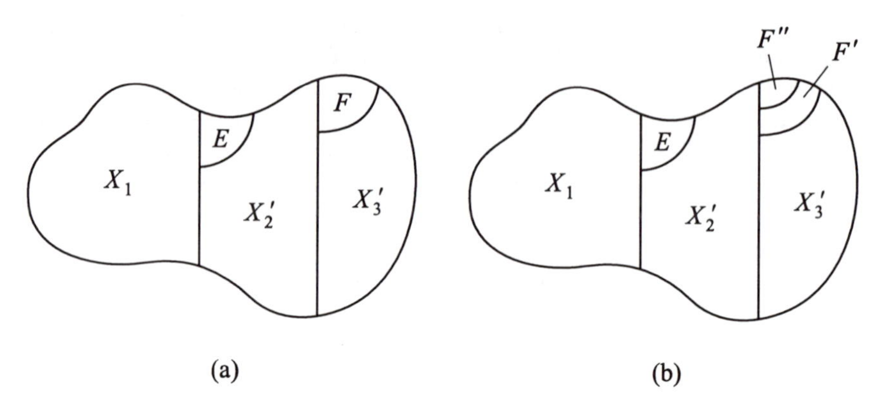

Figure 5.6.

Oskui Three Person Envy-Free Dirty Work Algorithm

We wish to assign Tom, Dick, and Harry envy-free portions of the lawn to mow. Not surprisingly, we ask Tom to cut the lawn in three equal parts, $X = X_1 \cup X_2 \cup X_3$. If Dick and Harry disagree on which is the smallest job, the envy-free assignments are easily made and we are done (another example of the serendipity of disagreement).

So we may assume in all that follows that Dick and Harry both view X_1 as the strictly smallest portion of lawn to mow. Now we ask Dick to cut $X_2 = X_2^{'} \cup E$ and $X_3 = X_3^{'} \cup F$ so that $\mu_D(X_1) = \mu_D(X_2^{'}) = \mu_D(X_3^{'})$ as shown in Figure 5.6(a).

Case 1: If Harry thinks both $X_2^{'}$ and $X_3^{'}$ are at least as large as X_1, then the three pieces $X_1, X_2^{'}$, and $X_3^{'}$ can be assigned envy-free. Let Tom choose first. We may assume without loss of generality that he chooses $X_2^{'}$, so $\mu_T(E) \geq \mu_T(F)$. Harry gets X_1 while Dick gets $X_3^{'}$.

Next, Dick cuts both E and F into three equal pieces, $E = E_1 \cup E_2 \cup E_3$ and $F = F_1 \cup F_2 \cup F_3$, and the three pieces of both E and F are chosen in the order of Harry, Tom, and Dick. We need only check that Tom does not envy Harry's job, since Harry chose first and Dick cut equal pieces. Note that since Tom gets at most $1/2(\mu_T(E) + \mu_T(F)) \leq \mu_T(E)$, Tom has a lawn to mow which is no bigger than X_2. But Harry has to mow at least all of X_1, and since Tom views X_1 and X_2 as equal tasks, Tom won't envy Harry and Case 1 is completed.

Case 2: On the other hand, Harry might think X_1 is at least as large as both $X_2^{'}$ and $X_3^{'}$. Then we could ask him to increase (if necessary) both $X_2^{'}$ and $X_3^{'}$

until he considers them both equal to X_1. But now Dick views X_1 as no bigger than (the new) X_2' and X_3'. With Dick and Harry reversing roles, we are back to Case 1.

Case 3: This leaves only the case where Harry considers X_1 to be neither the easiest nor the hardest job. We may assume $\mu_H(X_2') \geq \mu_H(X_1) \geq \mu_H(X_3')$ with at least one inequality strict. We now ask Harry to cut $F = F' \cup F''$ so that $\mu_H(X_3' \cup F') = \mu_H(X_1)$, as seen in Figure 5.6(b). This can be done since Harry considers X_1 no bigger than X_3. In this case Tom gets to choose first between X_2' and $X_3' \cup F'$, producing two subcases where the other pieces are assigned as summarized below. It can be checked that the previous inequalities ensure that the assignments are envy-free.

	Tom chooses	**Dick gets**	**Harry gets**	
Subcase 1:	X_2'	X_1	$X_3' \cup F'$	(here $\mu_T(E) \geq \mu_T(F''))$)
Subcase 2:	$X_3' \cup F'$	X_2'	X_1	(here $\mu_T(E) \leq \mu_T(F''))$)

In the first subcase E and F'' are both cut into three equal pieces by Harry, and the pieces of both E and F'' are chosen in the order Dick, Tom, and Harry. In the second subcase Dick cuts both E and F'' into three equal pieces, and the choosing is done in the order Harry, Tom, and Dick. The fact that the resulting division is envy-free is checked in a similar way as in Case 1 above. In both subcases Tom gets a job no worse than mowing X_1, and only whoever has been assigned X_1 chooses ahead of him.

In all cases the lawn is now ready to mow in an envy-free manner and the algorithm is complete.

Oskui also observes that the single Moving Knife Algorithm found in Exercise 5.4 can be easily altered to cover the dirty work problem. He also shows how to alter Stromquist's Four Moving Knives Algorithm to give a solution to the envy-free dirty work problem. These are left for the exercises.

5.6 Existence Theorems

We have referred to theorems guaranteeing the existence of pieces but not giving any method of finding them. When trying to find pieces of a cake to accomplish some sort of desired fair division, there are a couple of basic questions. First, quite apart from whether or not we know how to find them, do they exist? There is no use in looking for a general algorithm to ensure three persons get 40% of the cake by their assessments, because that algorithm would have to apply to the case where all three were in total agreement on all pieces. (For example, each may simply assign value by volume.) The pieces we would be looking for do not always exist.

If we know they exist in every case, can we actually find them? Not necessarily. There are many instances in mathematics in general, and cake-cutting in particular, where the existence of mathematical objects is guaranteed but there is no known way to produce them. Thus, we distinguish between *existence* proofs and *algorithmic* or *constructive proofs.* The first type simply convinces us that something exists with no method provided to produce that something. The second type not only convinces us they exist but does much more by actually producing the objects.

One of the first examples illustrating the distinction between these two types of proof which we encounter as we learn mathematics is in the solution of polynomial equations. Does every quadratic polynomial equation with complex coefficients have a complex root? Yes. How do you know? If the equation is $ax^2 + bx + c = 0$, the roots are $(-b \pm \sqrt{b^2 - 4ac})/2a$. We can easily produce them. How about cubic equations? Yes, they have roots and we can produce them using a half of a page of computations rather than the one line used for the quadratic. How about quartic equations? The answer is the same, but a page and a half of computation is required. How about degree five or more? The Fundamental Theorem of Algebra proves the existence of roots but doesn't show how to find them. Indeed, we know there cannot be a method similar to the ones used to produce roots for linear, quadratic, cubic, and quartic equations. The roots exist, but we generally have to settle for approximations to their actual values.

In cake cutting, existence proofs for certain types of fair division were given before algorithms were discovered to construct the necessary pieces. In the case of exact division, an existence proof says the required cake portions exist, but we still don't have any way of producing them [Alo], [DS], [Ste2], [Woo2]. Indeed, we will see that no *finite* algorithm can always produce the pieces (see Section 8.3). Since our primary focus is on algorithms that actually produce portions for various types of fair division, we will spend little time examining existence proofs.

5.7 Classes of Algorithms

We have encountered algorithms of entirely different characters which we have noted along the way. Let us review some things about these different classes of algorithms.

Moving Knife Algorithms, which require a continuum of decisions as time and knives move, we call *continuous* algorithms. These tend to be very powerful, work more problems, and get things done quickly with fewer cuts.

On the other hand, there are the finite algorithms which fall into two categories, *bounded* and *unbounded.* As we have said, we can promise that, even in the worst case, only three cuts will be required for the Trimming Algorithm to accomplish fair division for three persons. For the Successive Pairs Algorithm for three persons, five cuts will always suffice. If we have a finite algorithm to accomplish some task, and if we can know in advance that in every case some fixed number of cuts will always suffice, the finite algorithm is *bounded.*

On the other hand, Woodall's Algorithm will always accomplish strong fair division in some finite number of steps (and decisions), but we cannot say in advance it will all happen before 1,000 steps or 10,000,000 steps. We don't know how many steps may be required. It will depend on values of pieces that occur along the way, and we can't know in advance what those values will be. So although we know the algorithm will accomplish fair divison in a finite number of steps, we cannot give an upper bound on how many steps will be required. Such algorithms we call *finite unbounded.*

There is yet another class of algorithms, as we saw in Section 3.5, that are not continuous or finite. These we call *discrete infinite* algorithms. The algorithms will be described by some infinite sequence of steps, a first, then second, third, etc., that will accomplish the tasks. In these algorithms, no continuum of decisions is used, unlike Moving Knife Algorithms. Note that all finite algorithms are discrete because any finite number of tasks can be sequenced. Yet some algorithms whose steps can be sequenced nevertheless require an infinite number of steps.

5.8 EXERCISES

5.1. What does the implication diagram in Section 5.1 look like in the special case when $n = 2$?

5.2. Assume it is possible to give each of $P_1, \cdots, P_n$ a portion of cake each considers to have value strictly more than $1/n$. Show how (as in Section 5.2) to give each of $P_1, \cdots, P_{n+1}$ portions each considers strictly more than $1/(n+1)$.

5.3. Verify that the Stromquist Algorithm in Section 5.3 provides envy-free pieces for three players.

5.4. Show that the following Single Moving Knife Envy-Free Algorithm accomplishes envy-free division for three players.

Have a referee move a knife across the cake from left to right until the first player, say Tom, who feels a third of the cake is to the left of the knife, says "Cut."

Tom will get that piece, so set it aside. Have Tom rotate the knife continously over the remaining portion so that he thinks equal portions always lie on either side of the rotating knife. At some angle before the knife has made half a revolution, Dick will agree that the knife bisects this part of the cake (Why?), at which time Dick says "Cut." Harry gets his choice of these two pieces, Dick the other, and Tom gets the piece set aside. (What would happen if the knife were curved rather than straight?)

5.5. Fiona and Gwen could give information about their measures μ_F and μ_G, that is, information about how they value certain portions of the cake with *distribution functions* which we can describe in the following way. Assume the cake lies below the interval [0,1] on the x axis, and suppose Fiona's distribution function f has the property that if A_{ab} is the portion of cake lying between vertical lines $x = a$ and $x = b$ then $\mu_F(A_{ab})$ is the area bounded by $y = f(x), x = a, x = b$ and the x axis.[2]

Assume the function g has the similar property for Gwen. The graphs of f and g could look like this:

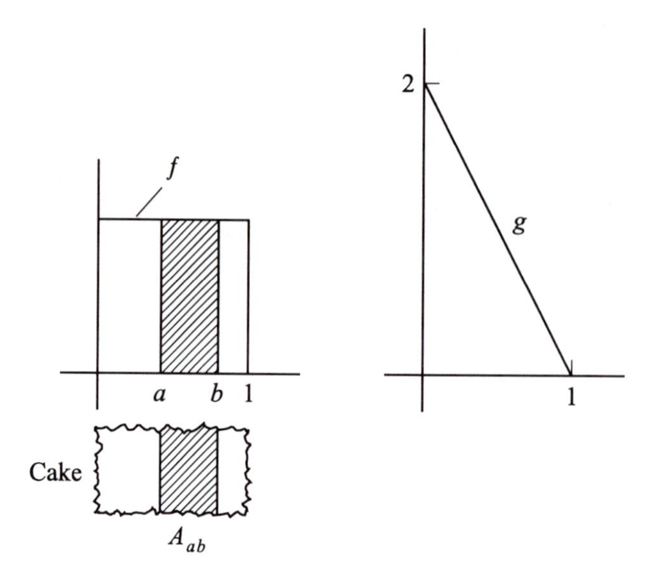

The shaded region has area $\mu_F(A_{ab})$.

[2]The function f can be obtained as follows (under suitable general assumptions about μ_F). Ask Fiona to define a function F whose domain is [0,1] by

$$F(x) = \mu_F(A_{0x}).$$

Then the distribution f is given by F'. Note that $F(0) = 0, F(1) = 1 = \mu_F(A_{01}) = \mu_F(X)$ and $F(a) = \int_0^a f(x)dx$ so that $\mu_F(A_{ab}) = \int_a^b f(x)dx$. The function F is called the *cumulative distribution function.*

We see in this case that Fiona thinks the value of the cake is *uniformly distributed* from left to right. If any two portions cut by vertical lines have the same width, they have the same value to Fiona. On the other hand, Gwen prefers the left side of the cake over the right side, because vertical slices under the graph of g have greater area near 0 than the same width slices near 1. Note that the area under both graphs on [0,1] is 1, since $\mu_F(X) = \mu_G(X) = 1$.

(a) For these two functions find a slice of cake between two vertical lines that both Fiona and Gwen value at 1/3 of the cake; i.e., find a and b so that $\mu_F(A_{ab}) = \mu_G(A_{ab}) = 1/3$.

(b) Argue that, for any value between 0 and 1 (not just 1/3), such a slice exists.

(c) Repeat part (b) for these two distribution functions.

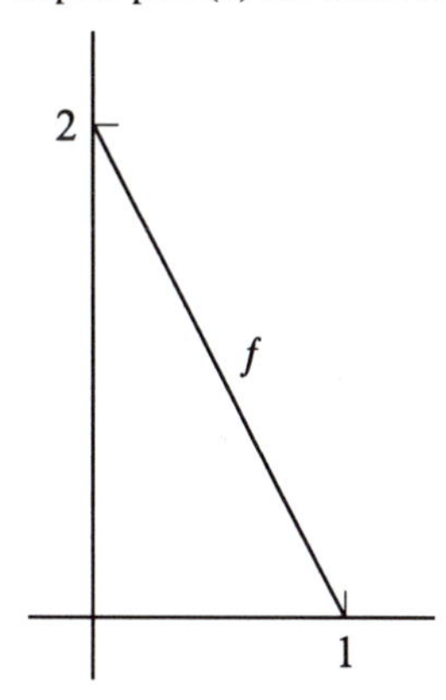

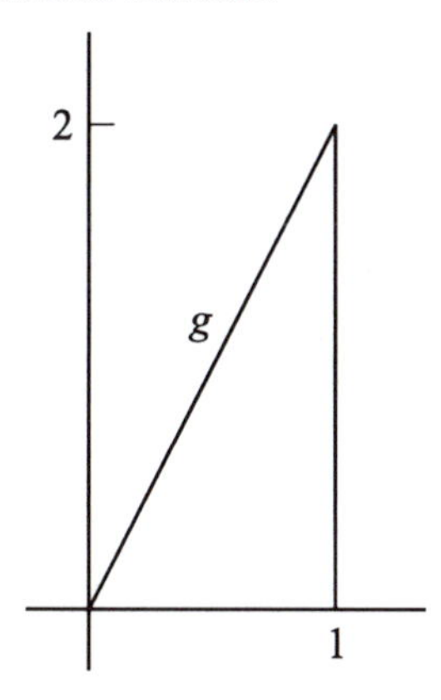

5.6. This exercise illustrates the Moving Knife Exact Division Algorithm of Section 5.4. Suppose the two distribution functions are as shown; Fiona values the cake uniformly, but Gwen doesn't like the center portion of the cake very much. (Again, the area under both graphs on [0,1] is 1.)

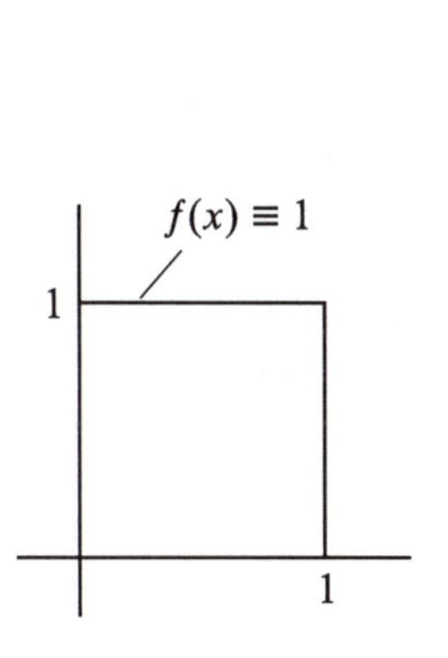

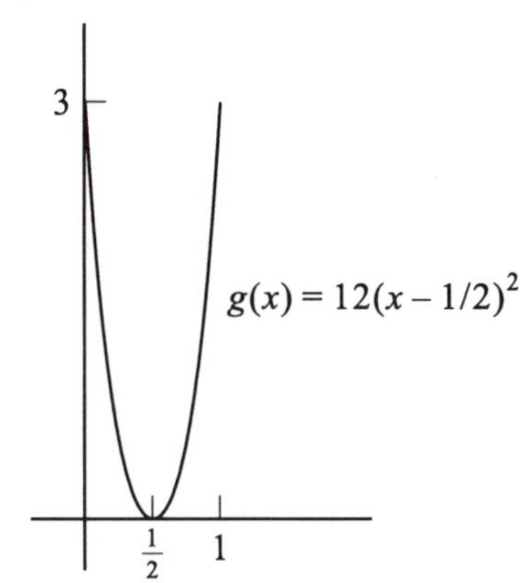

(a) Show that there is no slice which Fiona and Gwen will both assign value $3/4$; i.e., for no A_{ab} is $\mu_G(A_{ab}) = \mu_F(A_{ab}) = 3/4$. So as Fiona moves two vertical knives from left to right which always contain $3/4$ of the cake between them, there will be no time when Gwen will agree.

(b) Show that there is a slice they would both assign value $1/4$. How many such slices are there?

5.7. Show that if three pieces have been cut so that a simple fair division assignment is possible for three players, then the players can be ordered and fair pieces assigned so that the first player doesn't envy the other two , and the second player doesn't envy the third. (This result is generalized in Section 6.1.)

5.8. (a) Modify each of the following algorithms so as to apply to the dirty work problem:
(i) Trimming Algorithm,
(ii) Successive Pairs Algorithm,
(iii) The Kuhn Algorithm found in Exercise 1.8,
(iv) Cut Near-Halves Algorithm.

(b) In each case how does the number of cuts required compare to the number required by the original simple fair division algorithm?

5.9. If you are solving the dirty work problem for three players using the Trimming Algorithm found in Chapter 1, and if all three exercise the option to cut, can you give each a portion they consider strictly less than $1/3$?

5.10. Show how to modify the algorithm found in Exercise 5.4 so it solves the three person envy-free dirty work problem.

5.11. To modify Stromquist's Moving Knife Algorithm so that it applies to the dirty work problem, ask a referee to move a sword from left to right dividing X into two pieces, X_1 to the right of the sword and X_2 to the left. Simultaneously, have the three players adjust parallel knives over X_2 so that each of their knives cuts X_2 in exact halves. Instruct the players that when any one of them says "cut," the lawn will be cut in three portions by the referee's sword and the middle knife of the three players, creating partitions $X_2 = X_2^{'} \cup X_2^{''}$ and $X = X_2^{'} \cup X_2^{''} \cup X_1$. Further instruct them that they should say "cut" whenever they first think X_1 is smaller than or equal to both $X_2^{'}$ and $X_2^{''}$. Without loss of generality, we can assume that the players' knives from left to right belong to P_1, P_2, and P_3, respectively. Consider the three cases generated by which of the three players says "cut" and show how, in each case, to assign $X_2^{'}, X_2^{''}$, and X_1 to the players in an envy-free way.

5.12. What would prevent Tom, Dick, or Harry from waiting until the knife moves further to the right beyond where they were instructed to say "cut" in the Dirty Work Moving Knife Algorithm in Exercise 5.11?

5.13. Explore how to modify the method used to divide beachfront property in Section 4.2 so that it could be used for a dirty work problem. Show that if no two cuts coincide, all players can be given a task strictly smaller than the one they cut.

5.14. Give distribution functions $f(x, y)$ and $g(x, y)$ defined on the unit square $[0, 1] \times [0, 1]$ so that the two resulting measures assign the same value to any piece generated by a vertical cut, but assign different value to some pieces generated by horizontal cuts.

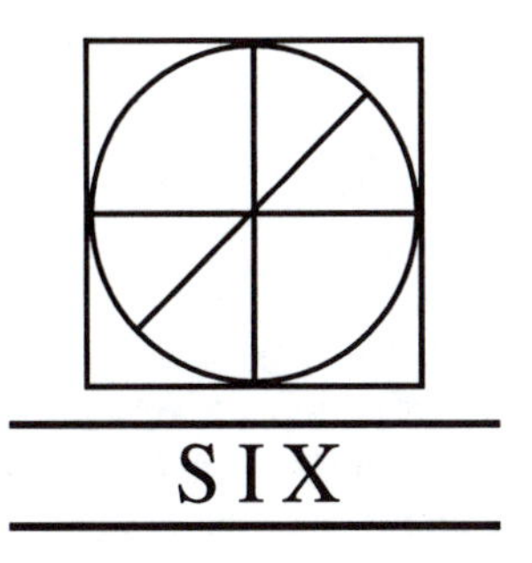

SIX

Some Combinatorial Observations

6.1 Applying Graph Theory to Fair Division

In this chapter we will see ways in which some questions about fair division can be explored in a graph theoretic setting. Readers wanting more detailed information about graphs can consult any standard book on this subject such as [CL], [Tuc].

Suppose there are n players $P_1, \cdots, P_n$ with measures $\mu_1, \cdots, \mu_n$ respectively and that the cake X has been somehow cut into m pieces $X_1, \cdots, X_m$. We will construct some graphs using $P_1, \cdots, P_n$ and $X_1, \cdots, X_m$ as our vertices. Depending on which version of fair division we are interested in, we draw an edge P_iX_j if and only if piece X_j is "acceptable" to player P_i. Thus, for simple fair division we draw P_iX_j if and only if $\mu_i(X_j) \geq 1/n$, whereas for envy-free division we draw P_iX_j if and only if $\mu_i(X_j) \geq \mu_i(X_k)$ for all k.

Example: The fairness graph below represents the facts that:
Pieces X_1 and X_2 are acceptable to P_1.
Pieces X_1 and X_3 are acceptable to P_2.
Piece X_2 is acceptable to P_3.

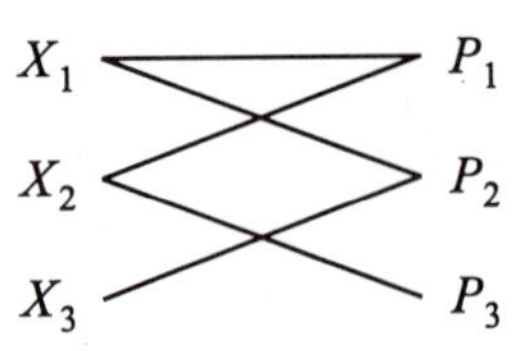

Notice that the vertices are naturally divided into the two subsets of pieces and players. Moreover, every edge joins a vertex in each subset. Such graphs are known as *bipartite* graphs, and they are the only kind we will use.

Looking at the example above, can we assign the pieces so that all of the players get an acceptable piece? It's easy to see that assigning X_1 to P_1, X_2 to P_3, and X_3 to P_2 will work. This assignment is represented by the subgraph:

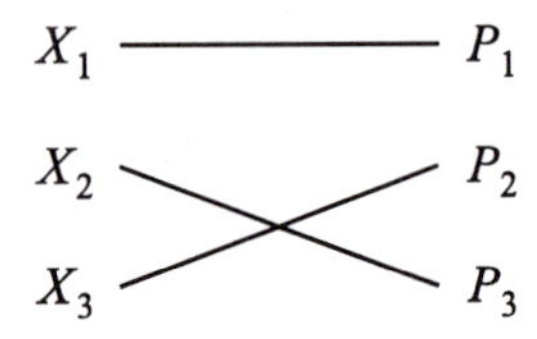

Suppose we have the same number of players and pieces. Any assignment of some of the pieces to the players such that each player gets at most one acceptable piece (even if not all of the pieces get assigned) is called a *matching*. If all of the pieces do get assigned, so necessarily all players get an acceptable piece, we have a *perfect matching*. So to accomplish a fair division one must find a perfect matching in the acceptability bipartite graph.

Obviously, not every fairness graph, constructed as above, has a perfect matching. Also, as the number of players and pieces increases it may not be easy to decide whether there is a perfect matching. A number of efficient algorithms exist for finding a largest possible matching in a graph. If such a matching uses all of the pieces, we have our desired perfect matching. These algorithms are beyond the scope of what we will need and can be found in books about combinatorial optimization [EM]. What we will use, however, is a very famous result, Hall's Theorem, which doesn't tell how to find a perfect matching, but does give a nice necessary and sufficient condition for one to exist.

Some notation is necessary. If we have a set of pieces S, we will need to count the number of players $N(S)$ who find one or more pieces of S acceptable. In the example above, if $S = \{X_1, X_2\}$, then $N(S) = 3$, since all three players like either X_1 or X_2. But since P_3 doesn't accept X_1 or X_3 for $S = \{X_1, X_3\}$, we have $N(S) = 2$. It seems reasonable that the larger these counts are for the various sets of pieces of cake, indicating there are more edges in the graph and more acceptable pieces, the better chance there is for a perfect matching. This is the intuitive idea that Hall's Theorem formalizes, which we now state for our cake-cutting setting.

Hall's Theorem. *A fairness graph for n players has a perfect matching if and only if, for every k with $1 \leq k \leq n$, $N(S_k) \geq k$ for every subset S_k having k pieces of cake.*

Returning to our example above, since there were only three players it was easy to give the subgraph which established the perfect matching. To apply Hall's Theorem, which would merely have established the *existence* of the matching, we would have to count $N(S)$ for all collections S containing one piece of cake (three of them), all collections containing two pieces (three of them) and all containing three pieces (one of them). For n players and pieces there are $2^n - 1$ such collections S to check — and that is a formidable job for large n. Fortunately, our application of Hall's Theorem will not require this to be done.

The condition $N(S_k) \geq k$ is clearly necessary for a perfect matching. If you have five pieces and only three players will accept some of them, it is certain in any assignment of pieces to players that at least two players will get unacceptable pieces. Showing the condition is sufficient is not so easy, but proofs can be found in most standard references on graph theory [CL], [Tuc].

Harold Kuhn gives an algorithm for simple fair division which generalizes what was started in Exercise 1.8 [Kuh]. The algorithm can also be found in the chapter by Kenneth Rebman in the book *Mathematical Plums* [Reb]. Recall the idea is to have one player, say P_1, cut the cake in n equal pieces and form an acceptability matrix A whose i, jth entry a_{ij} is given by

$$a_{ij} = \begin{cases} 1 \text{ if } \mu_i(X_j) \geq \dfrac{1}{n} \\ \\ 0 \text{ if } \mu_i(X_j) < \dfrac{1}{n}. \end{cases}$$

Note that all entries in the first row of A are ones, and each row has at least one entry which is a one. We hope to find an assignment of pieces to players so that each player gets a piece corresponding to an entry of one in matrix A. This isn't always possible (maybe all non-cutters will accept only X_1), and in this case we must give part of the cake to players willing to accept that part, while all the others are willing to give it away and divide the complementary portion of cake. Kuhn's Algorithm uses the Frobenius-König Theorem about permanents of $0 - 1$ matrices to accomplish this task.

The Matching Algorithm for Simple Fair Division

We now reformulate all of this in graph theoretic terms so we can apply Hall's Theorem to produce the Matching Algorithm. Note that the matrix A is just the *adjacency matrix* for our fairness bipartite graph; i.e., $a_{ij} = 1$ if and only if P_i and X_j are joined by an edge. We proceed by induction. For $n = 2$ players, the Matching Algorithm reduces to Cut and Choose, since the three possibilities for the fairness graph are

X_1 P_1 X_1 P_1 X_1 P_1

X_2 P_2 , X_2 P_2 , X_2 P_2 .

In any case, allowing P_2 to choose establishes the required matching.

So let us assume the Matching Algorithm does the job for $1, 2, \cdots, n-1$ players. We must now show how the algorithm handles n players. As before, P_1 cuts n equal pieces. If there is any piece X_i for which $N(\{X_i\}) = 1$, then this means that only P_1 accepts X_i, so give X_i to P_1 and let $P_2, \cdots, P_n$ divide the complementary set. They are happy to do this since all feel they got rid of P_1 by giving up an unacceptable piece X_i. The induction hypothesis applied to players $P_2, \cdots, P_n$ on that complementary portion completes the task when such an X_i exists.

So we next deal with the case where $N(\{X_i\}) > 1$ for all i with $1 \leq i \leq n$. The trick now is to find the smallest k, $1 < k \leq n$, for which there is some set S_k of k pieces of cake for which $N(S_k) = k$ and such that for any subsets $S_j \subset S_k$, $1 \leq j < k$, with j pieces, we have $N(S_j) > j$. Note that if the smallest k happens to be n, Hall's Theorem says a matching exists on the pieces already cut and we are done; all we have to do is find the matching. On the other hand, if $k < n$, we consider the subgraph whose vertices are the k pieces in S_k and the k players who accept some of those pieces. Hall's Theorem restricted to this subgraph insures a matching can be found assigning each of the k players an acceptable piece from S_k. The remaining $n - k$ players are happy to give those pieces away, since none of them are acceptable. So those $n - k$ players can happily divide the complementary portion, and we turn that task over to the induction hypothesis to get it done.

Hence, the algorithm is completely described when we know we can find that least k as just described. The existence of that k is assured by the following facts for the case we are considering:

(i) $N(S) \geq 2$ for any one-piece collection of pieces S.

(ii) $N(\{X_1, X_2, \cdots, X_n\}) = n$ since all players must like some piece.

(iii) If $N(S) \geq j$ for all collections S containing exactly $j - 1$ pieces, then $N(S) \geq j$ for all collections S containing exactly j pieces (since every collection having j pieces contains subcollections with $j - 1$ pieces).

We know from (i) that the k we seek is not $k = 1$, because $N(S)$ is always too big when S contains only one piece. Condition (iii) assures we won't move from any $j - 1$ where $N(S)$ is always too big to a j where some $N(S)$ is too small without finding some set S having j pieces with $N(S) = j$. Condition (ii) assures that if we don't find our desired k among the numbers $2, 3, \cdots, n - 1$ then $k = n$. This completes the description of the Matching Algorithm.

We can also use Hall's Theorem to prove the result promised in Exercise 5.7. If simple fair division is possible, then the players and pieces can be ordered so that P_i gets X_i, $\mu_i(X_i) \geq 1/n$ for all i, and no player envies a piece of any player further down in the ordering. Thus we can always guarantee a kind of *semi-envy*-free division.

Theorem 6.1. *If there is a simple fair division among players $P_1, \cdots, P_n$ using pieces $X_1, \cdots, X_n$, then the pieces can be reordered so that $\mu_i(X_i) \geq 1/n$ and $\mu_i(X_i) \geq \mu_i(X_j)$ for $1 \leq i \leq j \leq n$.*

Proof: We use induction in a way quite similar to the simple fair division algorithm above. Begin with the pieces $X_1, \cdots, X_n$ for which we now know simple fair division is possible and form the simple fairness graph G. If $n = 2$, then any simple fair assignment is also envy-free and hence semi-envy-free.

For $n > 2$, by Hall's Theorem we know $N(S_j) \geq j$ for all subsets of j pieces and all j, $1 \leq j \leq n$. Let k be the smallest integer such that $N(S_k) = k$ for some specific set S_k containing k pieces. If $k = n$, we can give P_1 his or her choice of pieces and know by Hall's Theorem that simple fair division is still possible among $P_2, \cdots, P_n$ using the $n - 1$ remaining pieces. Induction guarantees semi-envy-free division of these $n - 1$ pieces.

If $k < n$, again by Hall's Theorem, there is a perfect matching for the pieces in S_k and the k players who accept some piece in S_k. Let S_{n-k} be the pieces not in S_k. There must also be a matching of acceptable pieces from S_{n-k} with the remaining $n - k$ players, since such a reduced matching must have been a part of the given original matching of all n pieces with the n players. By induction both reduced matchings on S_k and S_{n-k} can be done so as to be semi-envy-free. Now simply use these two semi-envy-free matchings to order all players and their pieces by putting the players claiming pieces of S_{n-k} before those claiming pieces in S_k. Since players assigned pieces in S_{n-k} find pieces in S_k unacceptable, the overall ordering is semi-envy-free for all the players.

6.2 What Can Be Done with *n* Arbitrary Pieces?

One question we haven't addressed yet is what can be done if we are simply given the cake already cut up into n pieces $X_1, \cdots, X_n$. Clearly no simple fair division or envy-free division is always possible, since it might be the case that all of the players agree that only one of the pieces is acceptable. Indeed, they may all think that this one piece contains almost all of the value of the cake. We can't even specify any $\epsilon > 0$ and guarantee every player at least ϵ. Problems such as these are what motivate the search for fair division algorithms in the first place.

Since, in general, any kind of fair division we have previously discussed is going to be impossible, what can we do in an attempt at some type of reasonable distribution of the pieces? Because we are interested in everyone's evaluations of all the pieces, not just whether a piece is acceptable or not, we introduce the following representations of the evaluations.

The *value graph* is the complete bipartite graph where the vertices P_i and X_j are joined by an edge with value (or weight) $\mu_i(X_j)$. The *value matrix* $V = [v_{ij}]$ has the entries $v_{ij} = \mu_i(X_j)$.

Since it is impossible to guarantee a fair division in any of the senses previously discussed when we are presented with these n pieces already cut, is there any other natural condition to substitute for "fairness"? One good candidate is what is known as *Pareto optimality*. We should use a distribution of the pieces which has the property that no other distribution can give any player a larger share without giving a different player a smaller share.

Example: Consider the value matrix $\begin{bmatrix} .5 & .4 & .1 \\ .1 & .4 & .5 \\ .2 & .3 & .5 \end{bmatrix}$ and the three possible assignments indicated by **bold values**.

$$\begin{bmatrix} .5 & \mathbf{.4} & .1 \\ .1 & .4 & \mathbf{.5} \\ \mathbf{.2} & .3 & .5 \end{bmatrix} \qquad \begin{bmatrix} \mathbf{.5} & .4 & .1 \\ .1 & .4 & \mathbf{.5} \\ .2 & \mathbf{.3} & .5 \end{bmatrix} \qquad \begin{bmatrix} \mathbf{.5} & .4 & .1 \\ .1 & \mathbf{.4} & .5 \\ .2 & .3 & \mathbf{.5} \end{bmatrix}$$

I II III

Assignment I is not Pareto optimal since II is better for P_1 and P_3 and the same for P_2. That is, no one should object to moving from I to II and some player would strictly prefer II. But P_2 would object to a switch from I to III. Since there are only six possible ways to distribute the pieces, it is easy to check that assignments II and III are both Pareto optimal — some player would object to changing from either one.

Although Pareto optimal assignments are possibly the most natural and certainly the most studied, other criteria for the assignment of pieces to players can be given.

We will assume player P_i receives piece X_i, but we will allow the pieces to be relabelled. For example, if we switch pieces X_i and X_j so that X_i is relabelled as X_j and vice versa, columns i and j in the matrix V are switched. We will want to consider all possible permutations and relabellings of the pieces; i.e., we will want to consider all matrices that result from permuting columns of the matrix V.

A number of reasonable assignments might be nominated, including:

Condition 1.
Make Sure No Two Players Prefer to Trade Pieces
That is, choose a permutation so that $v_{ii} < v_{ij}$ and $v_{jj} < v_{ji}$ do not simultaneously hold for any i and j. If both inequalities do hold for i and j, then P_i and P_j should trade pieces — they both are happier and no other player is affected by the switch.

Condition 2.
Make Sure No Two Players Can Increase Their Combined Satisfaction by a Trade of Pieces
That is, choose a permutation so that $v_{ii} + v_{jj} \geq v_{ij} + v_{ji}$ for all i and j. If this inequality fails, P_i and P_j can trade pieces and increase their collective satisfaction, even though one of the two might lose value in the trade. No other players would be affected by the switch.

Condition 3.
Maximize the Collective Satisfaction
That is, maximize over all permutations of pieces the value $v_{11} + v_{22} + \cdots + v_{nn}$. If this is done, there can be no trading of pieces which increases the collective satisfaction $v_{11} + v_{22} + \cdots + v_{nn}$.

Condition 4.
Minimize the Worst Disappointment
That is, maximize over all permutations of pieces the value $\min\limits_{1 \leq i \leq n} \{v_{ii}\}$. This assignment takes the view that you wish to control the worst disappointment any player will suffer in receiving his or her piece.

Condition 5.
Maximize a Single Player's Satisfaction
That is, maximize over all permutations of pieces the value $\max\limits_{1 \leq i \leq n} \{v_{ii}\}$. This assignment ignores the collective interests of the players and just makes sure the piece liked best by any of the players goes to that player.

Do any of these conditions guarantee Pareto optimality? Condition 1 is essentially a pairwise Pareto optimality condition. Nevertheless, an assignment satisfying Condition 1 may not be Pareto optimal. The same is true for Condition 2 which is a pairwise version of Condition 3. (See Exercise 6.2.) Condition 3 does imply Pareto optimality, since increasing at least one v_{ii} while not decreasing any others will increase $v_{11} + v_{22} + \cdots + v_{nn}$.

Conditions 4 and 5 specify only the value of one piece, and so don't actually give a complete assignment of the pieces. In general, many assignments satisfy the given condition. Choose any one of these assignments and give the smallest or largest piece, respectively, to the player specified by the assignment. Remove this player and corresponding piece from consideration and repeat the process with the remaining players and pieces. Interpreted in this way, Conditions 4 and 5 give complete assignments for all of the pieces. These resulting assignments will be unique if all v_{ij} are distinct, but might not be unique in case some of the values are equal. It is not difficult to see that both of the extended conditions yield Pareto optimal solutions if all of the v_{ij} are distinct, but not necessarily if some equal values occur. (See Exercises 6.8 and 6.9.)

There is also the question whether efficient algorithms exist to produce distributions satisfying these criteria. The answer is "yes" and, without going into detail, the following comments give the rough idea.

Conditions 1 and 2 require checking whether switching pieces between various pairs of players gives a better distribution. The number of permutations is finite and the procedures cannot cycle back to a previous partition, since every switch produces an improvement in the appropriate measurement.

Condition 3 is probably the most sophisticated and hardest to implement. Nonetheless, it is equivalent to the well-known problem of finding a maximum weight matching in a bipartite graph. Various non-trivial algorithms are known to accomplish this task [EM].

Conditions 4 and 5 seem quite similar, but whereas Condition 5 is essentially trivial (just find the largest v_{ij} and give X_j to P_i), Condition 4 takes just a little thought. We don't want to use the smallest v_{ij} but instead want to eliminate it. If we do this successively, when do we stop? (See Exercise 6.10.)

6.3 EXERCISES

6.1. For a given permutation of the pieces, let us define the satisfaction S for that assignment by $S = v_{11} + \cdots + v_{nn}$.

(a) Verify that the switch indicated when either Condition 1 or 2 fails will increase S.

(b) Prove that for some permutation of pieces $S \geq 1$.

6.2. (a) Give an assignment which satisfies Condition 1 but is not Pareto optimal.

(b) Give an assignment which satisfies Condition 2 but is not Pareto optimal.

(c) Give an assignment which satisfies Condition 2 but not 3. (Hint: One example can be given for all three parts.)

6.3. Give an example of an assignment satisfying Condition 5 but not 3.

6.4. A "greedy" algorithm is one that makes an optimal immediate choice without regard to what may follow. In order to permute columns to maximize S (as in Exercise 6.1), we might try a greedy approach: First permute so that the largest entry in V is on the diagonal. Strike out that row and column and repeat.

(a) Give an example where this greedy approach does not maximize S.

(b) Show the greedy algorithm produces a matrix V with $S \geq 1$.

6.5. Is it always possible to permute columns in V so as to satisfy Condition 1? Condition 2?

6.6. From matrix V, we could define two related matrices: the **Envy Matrix** E has entries $e_{ij} = v_{ii} - v_{ij}$ (differences within rows); the **Disagreement Matrix** D has entries $d_{ij} = v_{jj} - v_{ij}$ (differences within columns).

(a) Verify the assignment is envy-free if and only if $e_{ij} \geq 0$ for all i, j.

(b) Give an example of an assignment which is not envy-free but with $d_{ij} \geq 0$ for all i, j.

(c) Show if $d_{ij} \geq 0$ for all i, j then V satisfies Condition 3. Show the condition is not necessary.

(d) Show that row sums in D are equal. What is that sum?

6.7. (a) If an assignment is simple fair, which of the five conditions are automatically satisfied?

(b) If an assignment is envy-free, which of the five conditions are automatically satisfied?

(c) Prove that the row sums in matrix D (see Exercise 6.6) are maximized by an envy-free assignment.

6.8. Show that the extended Condition 4 yields a Pareto optimal distribution if all of the values v_{ij} are distinct, but may not be Pareto optimal if an equal value occurs.

6.9. Repeat Exercise 6.8 using Condition 5.

6.10. If we try to satisfy Condition 4 by successively eliminating the smallest v_{ij}, is there an easy way to know when we have finally found the max $\min\{v_{ii}\}$ specified in Condition 4?

SEVEN

Interlude: An Inventory of Results

Before returning to some unfinished business we have left along the way in Chapters 1–6, it is appropriate that we summarize what has been accomplished so far. We have seen some different interpretations of "fair" when trying to fairly divide a cake among n players; we have seen different classes of algorithms that can be applied to the task; we have had some successes but have also met defeat along the way. The situation is summarized in the following charts, where the various algorithms are listed. As before, X is the cake, the players are $P_1, \cdots, P_n$, player P_i uses measure μ_i, X is partitioned $X = X_1 \cup \cdots \cup X_n$, and P_i receives piece X_i.

7.1 Algorithms for Fair Division

Definition of "Fair"	Status of Problem
Simple Fair $\mu_i(X_i) \geq 1/n$, $i = 1, 2, 3, \cdots, n$.	A number of algorithms are known for any number of players: • Trimming Algorithm (1.3, *finite bounded*) [Ste1]; • Moving Knife Algorithm (1.3, *continuous*) [Gar]; (cont.)

Definition of "Fair"	Status of Problem
Simple Fair (cont.)	• Successive Pairs Algorithm (1.3, *finite bounded*) [Saa]; • Kuhn and Matching Algorithms (Ex. 1.8 and also 6.1, *finite bounded*) [Kuh]; • Divide and Conquer Algorithm (2.4, *finite bounded*) [EP]; • Cut Your Own Piece (4.2 A, *finite bounded*) [Ste3].

Comments: Of all the general finite algorithms, the Divide and Conquer Algorithm uses the fewest number of cuts for n players; specifically, $1 + nk - 2^k$ cuts may be required where $k = \lfloor \log_2 n \rfloor$, [RW1]. *Ad hoc* variations for small n lead to minor improvements as seen in Section 9.4.

Strong Fair $\mu_i(X_i) > 1/n$, $i = 1, 2, \cdots, n$.	Given a piece on which there is disagreement, algorithms that provide strong fair division exist: • Woodall Algorithm (5.2, *finite unbounded*) [Woo2]. • Also see the discussion in 4.2 A and B.

Comments: This is possible only when there is disagreement. There is a proof for the *existence* of strongly fair sets when just two of the measures are different [DS], [Reb], but all known algorithms start with a given piece on which there is disagreement.

Envy-Free $\mu_i(X_i) \geq \mu_i(X_j)$ for all i, j.	A number of algorithms have been given: • For $n = 3$, Selfridge's Algorithm (1.4, *finite bounded*, uses at most 5 cuts) [Str], [Woo1]; • For $n = 3$, Stromquist's Moving Knife Algorithm (5.3, *continuous*) [Str]; • For $n = 3$, a Single Moving Knife Envy-Free Algorithm (Ex. 5.4, *continuous*); • For $n = 4$, Brams, Taylor, and Zwicker Moving Knife Algorithm (5.3, *continuous*) [BTZ2]; • For any n, Brams and Taylor Algorithm (seen later in 10.3, *finite unbounded*) [BT2]; • For any n, Envy-Free, Near-Exact Algorithm (seen later in 10.3, *finite unbounded*) [RW4]; • For $n = 3$, unequal rational shares (see algorithm in 11.4, *finite bounded*) [Web3].

Comments: The *existence* of sets providing envy-free division has been known for some time [Alo], [DS], [Ste2], [Str], [Woo1]. Only recently have algorithms been given for any number of players, but although *finite*, they are not *bounded*. It would be nice to know whether or not there are *finite bounded* algorithms, even for $n = 4$.

Definition of "Fair"	**Status of Problem**
Super Envy-Free $\mu_i(X_j) < 1/n$ for $i \neq j$.	As with strong fair, super envy-free division can be done if there are pieces showing sufficient disagreement: • Given pieces that guarantee that the measures are independent, an algorithm has been given (10.4, *finite unbounded*) [Web4].
Comments: An *existence* proof has been given which shows, for any n, that this is possible if and only if the n measures are linearly independent (5.3 and 10.4) [Barl].	
Exact $\mu_i(X_j) = 1/n$ for all i, j.	Exact division is very difficult and few algorithms are known: • For $n = 2$, there is a single Moving Knife Algorithm (5.3, *continuous*) and a two Moving Knives Algorithm (5.4, *continuous*).
Comments: There can be no *finite* algorithm, bounded or unbounded, even for two players (to be seen in 8.3 [RW4]); the *existence* of sets for exact division has been established [Alo], [DS], [Ste2], [Woo2]; a finite "near-exact" algorithm has been given (5.4 and 10.2) [RW4].	
Dirty Work Fair $\mu_i(X_i) \leq 1/n$ for $i = 1, 2, \cdots, n$.	Many algorithms for fair divison can be modified for dirty work: • In Exercise 5.8, the standard algorithms for simple fair division are modified for the dirty work problem.
Comments: Not much work has been done to date on the dirty work problem.	
Dirty Work Envy-Free $\mu_i(X_i) \leq \mu_i(X_j)$ for $i \neq j$.	Two algorithms have been given: • For $n = 3$, the Stromquist Envy-Free Moving Knife Algorithm can be modified for the dirty work problem (Ex. 5.11, *continuous*) as can another continuous algorithm (Ex. 5.10, *continuous*); • For $n = 3$, Oskui Envy-Free Dirty Work Algorithm (5.5, *finite bounded*, uses at most 9 cuts) [Osk].
Comments: Many questions are still open, for example, what can be done with more than three players, or for players with unequal shares.	

7.2 Number of Cuts for Fair Division

The issue of the number of cuts for simple fair divison was raised by Steinhaus when he introduced the problem in 1947, and the problem has received considerable attention over the years. One can pose the question:

> If n people are to divide the cake and if we are allowed only k cuts, $k \geq n - 1$, just how much cake can we guarantee each of the players?

The following table summarizes some things that are known for small numbers of players. The numbers displayed in bold type are known to be the most cake that can be guaranteed each player with the specified number of cuts.

	n players							
	2	3	4	5	6	7	8	
$(n-1)$ cuts	**1/2**	**1/4**	**1/6**	**1/8**	**1/10**	**1/12**	**1/14**	**1/(2n − 2)** continues
n cuts		**1/3**	**1/4**	**1/6**	**1/8**	**1/10**	**1/12**	**1/(2n − 4)** continues
$n+1$ cuts				**1/5**	1/7	1/9	1/11	1/(2n − 5) continues
$n+2$ cuts					1/6	1/8	1/10	

Clearly, if n players can be guaranteed at least $1/n$ using k cuts, then they can also get $1/n$ using more cuts, so these spaces have been left blank. The values listed in the first two rows are established in Section 9.3. Thus, we can guarantee all n players a share of size $1/(2n-2)$, and no more, if we are to produce the same number of pieces as players using a finite algorithm. A specific algorithm of Even and Paz found in Section 2.6 gives four players each 1/4 using at most 4 cuts.

The entries in Row 3 are established in Sections 9.2 and 9.3. Special *ad hoc* algorithms for entries in Row 4 are needed. (See [Web2].) We do not yet know whether 7 or 8 cuts are necessary to give 6 players all a share of size 1/6.

The best results known at this time for a *general* n use the inductive procedure of the Divide and Conquer Algorithm together with *ad hoc* arguments for certain small values. The number of cuts grows as $n \log_2 n$. The table seems to hint that $O(n)$ may suffice, but we may be merely observing Richard Guy's Law of Small Numbers [Guy] — we just haven't looked at large enough cases yet to see the true pattern. Determining the true growth rate for the minimum number of cuts for simple fair division remains one of the most intriguing unsolved problems in the subject of fair division. It may also be one of the hardest.

The table raises some interesting questions:

1. Are the best possible entries always Egyptian fractions?

2. If the answer to Question 1 is "yes," do denominators in rows always increase by no more than two?

3. Are columns strictly monotone? (Even this most basic question is apparently still open.)

We note that if the answer to all three is "yes," then $O(n)$ cuts suffice for simple fair division. Yet we have been conditioned to expect that $O(n \log_2 n)$ may be the best possible bound.

EIGHT

Impossibility Theorems

8.1 Resetting the Stage

At various places, we have stated that no algorithm of a certain type can be found to accomplish a given task. For example, a *finite* algorithm cannot accomplish fair division for three players using only two cuts, and no finite algorithm can accomplish equal divisions. We return now to these issues and justify some of the claims made. *All algorithms considered in this chapter will be finite.*

If one wants to show that something *can* be done, simply exhibiting a correct algorithm to accomplish the task settles the issue. To claim that something *can't* be done (such as giving each of three players 1/3 of the cake with a finite algorithm using fewer than 3 cuts) is a different and more difficult matter. We must show that, whichever algorithm is used from the class of algorithms being considered as candidates to accomplish the task, none will always produce the desired division. For this reason, we need to state carefully the assumptions we are making about these algorithms. What are the rules of the game?

All of our impossibility theorems rely on one key fact that we have used throughout: *Any cut that is made can produce pieces of specified sizes for the cutter only, and any non-cutter can view the two new pieces as having any values compatible with that player's value of the piece before it was cut.* It will be assumed that the cake is a compact subset of E^n with positive Lebesgue measure and that all pieces cut are also Lebesgue measurable. We further assume that any measure μ used by any player is absolutely continuous with respect to Lebesgue measure. Thus, each Lebesgue measurable set is also μ measurable, and $\mu(A) = 0$ if A has Lebesgue measure zero. (For example, this implies for the Moving Knife Algorithm that it doesn't matter who gets the cake on the cut line since all players judge it to have value zero.) The converse is not assumed — for example, a player who does not like white cake may assign a piece value zero even though the piece has positive Lebesgue measure. We will assume that at each stage the algorithm will call upon one of the players to cut an existing piece and can specify what size the two new pieces are to be in the cutter's measure. (For whatever reason the cutter may not cut the specified sizes, so our impossibility arguments will not assume pieces of these sizes. See the discussion in Section 2.1 where this issue is addressed.) This next cut may depend on the values all players place on all existing pieces.

We assume there is no consultation among the players before the cuts. Suppose we ask Tom to cut an existing piece into what he considers halves, which Dick would judge to be in the ratio of 2 : 1. It is clear this couldn't always be done even if Tom had full information on both measures. If we assumed that every player had full *a priori* knowledge of the measure of every other player, the context of the problem would change from a combinatorial/game theory setting to a more measure theoretic setting in which stategic cutting might add another dimension to the problem.

Informally, we will think of the algorithm as a branching tree, rooted at the top by the cake X on which all players place value 1. Someone specified by the algorithm makes a cut, and at the next level of vertices are the two pieces produced by the cut $X = X_1 \cup X_2$, along with the values that the players place on these pieces. The algorithm must accommodate any evaluations compatible with the additivity requirement that $\mu(X_1) + \mu(X_2) = \mu(X)$. Branching is produced by different evaluations on the pieces by the players (i.e., by different measures used by the players) and by different subsequent cuts called for by the algorithm. Those cuts are based on the existing pieces and the values placed on them by the players. Even the two pieces X_1 and X_2 are not unique, since the cutter may be able to cut the required pieces in many different ways.

An algorithm is *finite* if every path in the tree terminates, which happens when a vertex is reached where the pieces can be partitioned into n subsets and assigned to the n players with each player's assessment of his or her portion providing the prescribed fair share. The algorithm is *bounded* if every path terminates within a *fixed number* of steps.

To illustrate, let us look at the simplest of all algorithms, Cut and Choose, which terminates after one step (or cut):

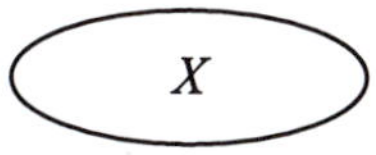

X

$$\mu_1(X) = \mu_2(X) = 1$$

Step 1: P_1 cuts halves

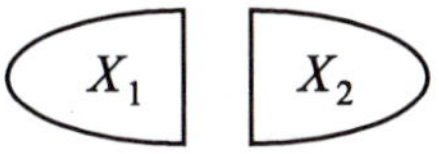

X_1 X_2	X_1 X_2
$\mu_1(X_1) = \mu_1(X_2) = 1/2$	$\mu_1(X_1) = \mu_1(X_2) = 1/2$
$\mu_2(X_1) > 1/2$	$\mu_2(X_1) \leq 1/2$
Algorithm terminates:	Algorithm terminates:
P_2 gets X_1, P_1 gets X_2	P_1 gets X_1, P_2 gets X_2.

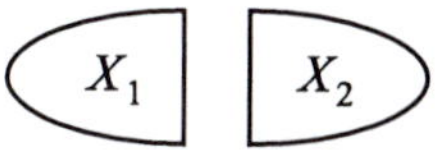

We have drawn only two branches because we can group cases where $\mu_2(X_1) \leq 1/2$ and cases where $\mu_2(X_1) > 1/2$. In general there may be many branches, perhaps infinitely many, depending on the evaluations (see for example the algorithms in Section 10.3). The algorithm can give different instructions for the next step, depending on the branch followed. We will assume every step consists of cutting one existing piece into two new ones. This can accommodate instructions such as, "cut this piece into equal thirds," by which we mean that the cutter cuts in the ratio $1 : 2$ as a first step and then cuts the larger (to the cutter) piece into equal parts as the next step.

To summarize, a finite algorithm is understood to have the following properties:

1. The algorithm starts with a cake X that is a Lebesgue measurable subset of E^n for which $\mu_i(X) = 1$ for all players P_i.

2. At each step, some player, say P_i, will cut an existing piece A into Lebesgue measurable pieces, $A = A_1 \cup A_2$, of specified size satisfying $\mu_i(A_1) + \mu_i(A_2) = \mu_i(A)$. For any $j \neq i$ and any two non-negative numbers satisfying $a_1 + a_2 = \mu_j(A)$, it may be that $\mu_j(A_1) = a_1$ and $\mu_j(A_2) = a_2$.

3. After a cut is made, all players make known their evaluation of the new pieces. On the basis of these values and all previously known values, the algorithm specifies the next cut.

4. The above procedures are repeated a finite number of times and the resulting pieces distributed to the players to accomplish the fair division.

With these assumptions about the algorithms we can now identify certain tasks that none of them will accomplish.

8.2 More Than $n - 1$ Cuts Are Required for n Players, $n \geq 3$

We know that using Cut and Choose two players can each receive at least half of the cake using only one cut, so that each player receives a portion consisting of a single piece. In Section 2.5 we saw that if three players use only two cuts, the most that all can be guaranteed is 1/4 of the cake. Not surprisingly this result generalizes to more players. In this section we show that simple fair division for n players cannot be accomplished by a finite algorithm using only $n - 1$ cuts. Later, in Chapter 9, we will return to the general question of what can be accomplished if the number of cuts is limited.

We first prove a Lemma.

Lemma 8.1. *Suppose a piece of cake A is to be divided by a finite algorithm into smaller pieces with players P_i and P_j each receiving a single subpiece of A. It is impossible to guarantee both P_i a piece whose measure exceeds $\mu_i(A)/2$ and P_j a piece of positive measure.*

Proof: If someone other than P_i makes the first cut on A, say $A = B \cup C$, it may be that $\mu_i(B) = \mu_i(C) = \mu_i(A)/2$. Therefore, P_i must make the first cut to receive more than $\mu_i(A)/2$ in a single piece. If P_i cuts $A = B \cup C$ with $\mu_i(B) > \mu_i(A)/2$, it may be that $\mu_j(C) = 0$. Thus, P_i and P_j must now each receive one subpiece from B, and we are essentially where we were at the outset, except that piece A has been replaced by B. We now must divide B so as to give the required portions. This process could continue indefinitely and thus not produce the required pieces after any finite number of steps. So the required finite algorithm is impossible.

We comment that the sequence of cuts referred to in Lemma 8.1 may appear to be rather pathological. If players use two independent measures, how likely is it that the second player will always view the piece offered as entirely worthless? In fact, this could naturally happen in actual cake-cutting situations. The cake may be chocolate on the left and white on the right. Player P_i, who must receive what is considered to be more than half of the cake, may repeatedly cut so that all of the chocolate portion lies in piece B. But player P_j may not like white cake and just as soon eat sawdust. So P_j views every piece offered as worthless, even though player P_i may feel pieces of positive value are being offered.

This lemma can be used to give a short proof of the result found in Section 2.5.

Theorem 8.1. *No finite algorithm can guarantee each of three players at least a third of the cake using only two cuts.*

Proof: Someone, say P_3, cuts $X = X_1 \cup X_2$ first. It may be that $\mu_1(X_1) = \mu_1(X_2) = \mu_2(X_1) = \mu_2(X_2) = 1/2$. It can be assumed without loss of generality that P_1 must share X_1 with some other player so that each receives a piece they value as at least 1/3. This is impossible by Lemma 8.1.

More generally we can prove:

Theorem 8.2. *No finite algorithm can guarantee each of n players at least $1/n$ of the cake using only $n-1$ cuts when $n \geq 3$.*

Proof: The case $n = 3$ is covered in Theorem 8.1, so assume $n \geq 4$. Someone, say P_n, makes a first cut $X = X_1 \cup X_2$. It may be that $\mu_i(X_2) = (1/n) - \epsilon$ whenever $1 \leq i \leq n-1$, where $\epsilon = 3/n(n-3)(2n-1)$. Now $P_1, \cdots, P_{n-1}$ must all receive a piece of X_1 they each consider to have value at least $1/n$. Someone, either P_n again or (without loss of generality) P_{n-1}, cuts $X_1 = Y_1 \cup Y_2$ and it may be that $\mu_i(Y_2) = (1/n) - \epsilon$ whenever $1 \leq i \leq n-2$. Now, $P_1, \cdots, P_{n-2}$ must each receive a piece of Y_1 they each consider to have value at least $1/n$. Continuing in this way, we eventually reach a point where P_1, P_2, P_3 must share a piece that is worth $3/n + (n-3)\epsilon = 6/(2n-1)$ to each of them. Suppose (without loss of generality) that P_3 cuts and both P_1 and P_2 value both pieces at $3/(2n-1)$. One of them must share one of the two pieces with another player so that each receives at least $1/n$. Since $2/n > 3/(2n-1)$ for $n > 3$, Lemma 8.1 shows the required division is impossible.

8.3 No Finite Algorithm Can Accomplish Exact Division

Given any ratio $a : b$ (rational or irrational), in Section 5.4 a Moving Knife Algorithm was described producing portions that each of two players agreed were in the exact ratio $a : b$. We next show that no finite algorithm, bounded or unbounded, can accomplish that task. We will prove the result for the simplest of all ratios $1 : 1$. We cannot know which algorithm is being considered, but regardless which it is, it will have to accomplish the task when the two players are using any probability measures μ_1 and μ_2 on X that are absolutely continuous with respect to Lebesgue measure. Information about the two measures is divulged only as the cuts are made and the resulting pieces evaluated. The plan will be to trace an infinite path through the algorithm tree. This will be done by showing that for any number of steps there is a branch of the algorithm that does not terminate at this level. Moving to the next level requires another cut, and we show how certain evaluations at this next level can also produce a non-terminating vertex. Arranging these steps sequentially produces an infinite path in the tree, i.e., a case for which the algorithm fails to terminate in any finite number of steps. So the essence of our task is to show how to move from one non-terminating vertex in the tree produced by the given algorithm to a next vertex where the algorithm also fails to terminate [RW4].

Theorem 8.3. *There is no finite algorithm that accomplishes exact fair division for two players in the ratio* $1:1$.

Proof: There must certainly be at least two pieces. At Step 1, even if the cutter produces exact halves, there are measures the non-cutter may be using that do not agree with the cutter's evaluations. This possibility creates a level two vertex where the algorithm does not terminate.

Proceeding by induction, we assume that after k steps we have arrived at a non-terminating case (vertex) where pieces $A_{1,k}, A_{2,k}, \cdots, A_{k,k}$ have been cut and values $a_{i,j} = \mu_1(A_{i,j})$ and $b_{i,j} = \mu_2(A_{i,j})$ have been provided by the two players. (This is all that is known at this stage about μ_1 and μ_2.) Based on this information, the algorithm prescribes which piece is to be cut next, who is to cut, and the sizes of the pieces the cutter should produce.

By renumbering if necessary, we may assume that the piece to be cut is $A_{k,k}$. This means that $a_{i,k} = a_{i,k+1}$ and $b_{i,k} = b_{i,k+1}$ for all i with $1 \leq i \leq k-1$. Let us assume the algorithm calls for P_2 to cut $A_{k,k} = A_{k,k+1} \cup A_{k+1,k+1}$ with specified values $b_{k,k+1}$ and $b_{k+1,k+1}$ respectively, where $b_{k,k+1} + b_{k+1,k+1} = b_{k,k}$. We may further assume neither $b_{k,k+1}$ nor $b_{k+1,k+1}$ is zero, because if either were, P_1 could agree with that evaluation and nothing has been changed by the cut.

We now show two values that $a_{k,k+1}$ and $a_{k+1,k+1}$ may take so as to produce another non-terminating vertex. (It will be the case that neither will be zero, so our unending chain of steps will contain only sets of positive measure to both of the players.) Since the algorithm did not stop with pieces $A_{1,k}, \cdots, A_{k,k}$ we know there is no subset $S \subseteq \{1, 2, \cdots, k\}$ for which

$$\sum_{i \in S} a_{i,k} = 1/2 = \sum_{i \in S} b_{i,k} \ .$$

If $T \subseteq \{1, 2, \cdots, k-1\}$, the sum $\sum_{i \in T} a_{i,k}$ may or may not equal 1/2. Set $\eta = \min |1/2 - \sum_{i \in T} a_{i,k}|$ where the minimum is taken over all T for which the sum is not 1/2. (For $k = 2$ let $\eta = 1/2$.) Note that $\eta > 0$. Regardless of how the algorithm prescribes $b_{k,k+1}$ and $b_{k+1,k+1}$ player P_1 may judge the cut so that $a_{k,k+1} = \frac{1}{2} \min\{a_{k,k}, \ \eta\}$.

To verify that this evaluation by P_1 produces another non-terminating case, we observe that if $S \subseteq \{1, 2, \cdots, k+1\}$ with

$$\sum_{i \notin S} a_{i,k+1} = \sum_{i \in S} a_{i,k+1} = 1/2 = \sum_{i \in S} b_{i,k+1} = \sum_{i \notin S} b_{i,k+1}$$

we may assume $k \in S$ and $k+1 \notin S$, since the process did not terminate at the previous step. But then $\sum_{i \in S} a_{i,k+1} = 1/2$ is impossible because of the way η was chosen. If the sum without $a_{k,k+1}$ is 1/2, it is not 1/2 with $a_{k,k+1}$ since that term is non-zero. If the sum without $a_{k,k+1}$ is not 1/2, the term $a_{k,k+1}$ can make up at most half of the difference, and the sum misses 1/2 by at least $\eta/2$.

This produces the required next non-terminating case and the argument is complete.

8.4 PROJECTS

8.1. Supply the details for Theorem 8.3 when the ratio is not 1:1. In this case one can choose η to be linearly independent (over the field of rational numbers) of all previous values.

For the case where there are $n > 2$ players, exact division in some prescribed ratios $a_1 : a_2 : \cdots : a_n$ would provide exact division in the ratio $(a_1 + \cdots + a_{n-1}) : a_n$ in the case $\mu_1 = \mu_2 = \cdots = \mu_{n-1}$.

8.2. Given a piece on which there is disagreement, Woodall's Algorithm provides a finite *unbounded* algorithm for strong fair division. For envy-free division with $n = 3$, the Selfridge Algorithm is finite and *bounded.* For $n > 3$ two envy-free algorithms are given in Chapter 10, but both are finite *unbounded.* These algorithms also provide a method of accomplishing simple fair division among two players in unequal irrational ratios. We do not know of any finite *bounded* algorithms for this task.

This result leaves open the question of whether or not the unbounded algorithms can be replaced by bounded ones. We know of no situation where it has been proved that something can be accomplished with a finite unbounded algorithm yet cannot be done with a finite bounded one.

More specifically, it would be nice to know what the case is for envy-free division for four players. Find a bounded finite algorithm to accomplish the task or show an unbounded finite algorithm is the best we can do. Also, can we accomplish simple fair division in an irrational ratio such as $1/\pi : (\pi - 1)/\pi$ with a finite bounded algorithm, or is that impossible?

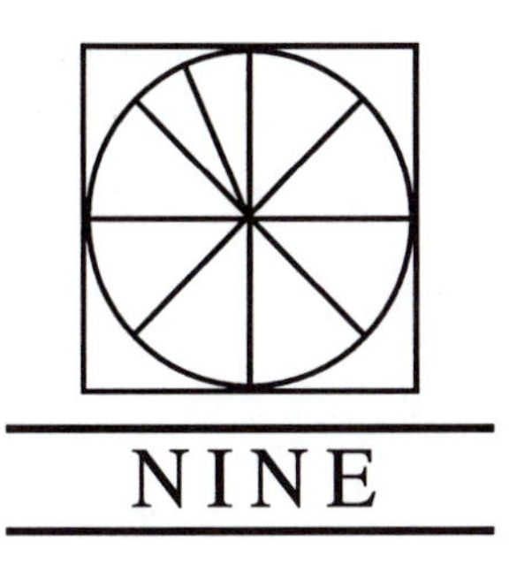

NINE

Attempting Fair Division with a Limited Number of Cuts

9.1 The Problem

Suppose n players are to share a cake and it is decided, for whatever reason, that only $n-1$ cuts can be used. Then each player will be given a portion of the cake consisting of a single piece. If fewer than $n-1$ cuts are used, we don't even have n pieces required to give each player a piece. We know from Section 8.2 that for $n \geq 3$ no finite algorithm can guarantee each player $1/n$ of the cake using only $n-1$ cuts. So how should the allotted cuts be utilized?

One could first take the point of view that we should give as many players as possible their $1/n$, knowing in advance that won't be everyone, and not worry what the others get. A second goal might be to try to give as much as we possibly can to every player, knowing it won't be $1/n$. In the later case, using the allotted number of cuts, we would try to maximize the value $\min_{1\leq i\leq n} \mu_i(X_i)$ in the same spirit as Condition 4 described in Section 6.2.

Presumably, a more satisfactory division in either case could be made if more cuts were allowed. So what could be done with n cuts (where one player would receive two pieces) or $n+1$ cuts? To define our problem more formally, suppose

n players are to share a cake X using k or fewer cuts. Motivated by the two possible points of view mentioned above, let us examine the two values

$$\begin{aligned} N(n,k) &= \max_{\mathcal{A}}(\text{number of } i \text{ such that } \mu_i(X_i) \geq 1/n), \\ M(n,k) &= \max_{\mathcal{A}} \min_{i}(\mu_i(X_i)), \end{aligned}$$

where $\mathcal{A}$ is the class of finite algorithms using at most k cuts which satisfy the conditions set down in Section 8.1. We will present some results, but there is yet no adequate resolution of all of the questions that come to mind.

9.2 Simple Fair Division for a Small Number of Players

We have seen finite algorithms that accomplish simple fair division for three and four players using three cuts (Section 2.3) and four cuts (Section 2.6) respectively. Also, in Section 8.2 we saw that for $n \geq 3$, $n-1$ cuts aren't enough to accomplish fair division for n players. What do we know for other numbers of players? Although the problem of determining the least number of cuts, $F(n)$, needed to guarantee each of n players a piece of size $1/n$ using a finite algorithm seems very difficult, perhaps we can make some progress when the number of players is small.

The first unresolved case is five players. The following algorithm shows that $F(5) \leq 6$. It will follow from Theorem 9.1 that in fact $F(5) = 6$.

Five-Player Six-Cut Algorithm

Proceed as follows:

Cut 1: P_1 cuts X into pieces A and B of size $3/5$ and $2/5$ respectively, producing the table of values:

	$A(3/5)$	$B(2/5)$
P_1	$=$	$=$
P_2	a_2	b_2
P_3	a_3	b_3
P_4	a_4	b_4
P_5	a_5	b_5

Table 9.1.

In the table the two equal signs signify P_1 accepts the two pieces A and B as having values exactly equal to their "target" values described by the algorithm.

For i with $2 \leq i \leq 5$, $a_i = \mu_i(A)$ while $b_i = \mu_i(B)$. Let us further agree to replace a_i by "+" (or "−") if $a_i \geq 3/5$ (or $a_i \leq 3/5$) and replace b_i by "+" ("−") if $b_i \geq 2/5$ ($b_i \leq 2/5$). That is, a "+" indicates the player accepts the piece as having value at least the target value, whereas a "−" indicates the player does not think the piece has more than the target value. The algorithm proceeds based on how many times "+" appears under piece A, and we treat the simpler cases first.

Case 1: Exactly three non-cutters accept A and Table 9.1 has (without loss of generality) the form:

	$A(3/5)$	$B(2/5)$
P_1	$=$	$=$
P_2	$+$	$-$
P_3	$+$	$-$
P_4	$+$	$-$
P_5	$-$	$+$

We now give piece A to players 2, 3, and 4 and give piece B to players 1 and 5 to divide equally in both situations. Since $F(2) = 1$ and $F(3) = 3$, at most $1 + 1 + 3 = 5$ cuts are used.

Case 2: Exactly two non-cutters accept A. Assume P_2 and P_3 accept A while P_4 and P_5 reject A and thus accept B.

Let players 1, 2 and 3 divide A while players 4 and 5 divide B. Again at most 5 cuts are used.

Case 3: If all four non-cutters accept A, Table 9.1 takes the form:

	$A(3/5)$	$B(2/5)$
P_1	$=$	$=$
P_2	$+$	$-$
P_3	$+$	$-$
P_4	$+$	$-$
P_5	$+$	$-$

Cut 2: P_1 cuts $B = C \cup D$ with $\mu_1(C) = \mu_1(D) = 1/5$.

Subcase i: Suppose evaluations of A, C, and D produce a table with at least one "+" in either of the last two columns so that we have (without loss of generality):

	$A(3/5)$	$C(1/5)$	$D(1/5)$
P_1	$=$	$=$	$=$
P_2	$+$	$+$	d_2
P_3	$+$	c_3	d_3
P_4	$+$	c_4	d_4
P_5	$+$	c_5	d_5

Now we can let P_1 have D, P_2 have C, and $P_3 - P_5$ divide A, using a total of at most 5 cuts.

Subcase ii: Suppose we have:

	$A(3/5)$	$C(1/5)$	$D(1/5)$
P_1	$=$	$=$	$=$
P_2	$+$	$-$	$-$
P_3	$+$	$-$	$-$
P_4	$+$	$-$	$-$
P_5	$+$	$-$	$-$

Let P_1 have C and let Players $P_2 - P_5$ divide $A \cup D$ using a total of at most 6 cuts, since $F(4) = 4$.

This completes Case 3 and only two cases, the most difficult to handle, remain. We treat them simultaneously since the strategy is similar in either case.

Case 4: If at most one non-cutter accepts A, then Table 9.1 takes one of the forms:

	$A(3/5)$	$B(2/5)$
P_1	$=$	$=$
P_2	$+$	$-$
P_3	$-$	$+$
P_4	$-$	$+$
P_5	$-$	$+$

or

	$A(3/5)$	$B(2/5)$
P_1	$=$	$=$
P_2	$-$	$+$
P_3	$-$	$+$
P_4	$-$	$+$
P_5	$-$	$+$

Cut 2: P_1 cuts A into pieces E and F of sizes $2/5$ and $1/5$ respectively, producing tables:

	$B(2/5)$	$E(2/5)$	$F(1/5)$
P_1	$=$	$=$	$=$
P_2	$-$	e_2	f_2
P_3	$+$	e_3	f_3
P_4	$+$	e_4	f_4
P_5	$+$	e_5	f_5

or

	$B(2/5)$	$E(2/5)$	$F(1/5)$
P_1	$=$	$=$	$=$
P_2	$+$	e_2	f_2
P_3	$+$	e_3	f_3
P_4	$+$	e_4	f_4
P_5	$+$	e_5	f_5

If all f_i are "–" (in either table) give F to P_1, and using at most 4 additional cuts let $P_2 - P_5$ divide $B \cup E$. If exactly one f_i, say f_2, is "+" give F to P_2, and let P_1, P_3, P_4 and P_5 divide $B \cup E$. Again at most 6 cuts total are used.

So we may assume (without loss of generality) that f_2 and f_3 are both "+" (in either table). In this case we ask P_1 to cut a third time.

Cut 3: P_1 cuts $E = G \cup H$ with $\mu_1(G) = \mu_1(H) = 1/5$. The two tables under consideration are now:

	$B(2/5)$	$F(1/5)$	$G(1/5)$	$H(1/5)$
P_1	$=$	$=$	$=$	$=$
P_2	$-$	$+$	g_2	h_2
P_3	$+$	$+$	g_3	h_3
P_4	$+$	f_4	g_4	h_4
P_5	$+$	f_5	g_5	h_5

or

	$B(2/5)$	$F(1/5)$	$G(1/5)$	$H(1/5)$
P_1	$=$	$=$	$=$	$=$
P_2	$+$	$+$	g_2	h_2
P_3	$+$	$+$	g_3	h_3
P_4	$+$	f_4	g_4	h_4
P_5	$+$	f_5	g_5	h_5

Each player must either accept $B \cup G$ at value $3/5$ or the complementary portion $F \cup H$ at value $2/5$. We proceed from these last two tables based on how many players accept $B \cup G$ or $F \cup H$.

Subcase 4-i: If at least two of $P_2 - P_5$ accept $F \cup H$, give $F \cup H$ to them. If possible include P_2 as one of the two. Give G to P_1 and B to the remaining two players. In the single exceptional case where P_2 accepts neither $F \cup H$ nor B, P_2 can be given piece G and P_1 is now one of the players sharing B. At most 5 cuts are used.

Subcase 4-ii: If exactly one of $P_2 - P_5$ accepts $F \cup H$, that player must accept either F or H, P_1 can be given the other of F or H, and $B \cup G$ is given to the other three players. At most 6 cuts are used.

Subcase 4-iii: If none of $P_2 - P_5$ accepts $F \cup H$, then all must accept $B \cup G$. Then give F to P_2, give H to P_1, and give $B \cup G$ to the other 3. At most 6 cuts are used.

This completes the algorithm showing that $F(5) \leq 6$, since at most 6 cuts were used in any case.

This algorithm shows that 6 cuts suffice for 5 players. We have seen earlier (Theorem 8.2) that $n - 1$ cuts suffice only when $n = 2$. That n cuts will satisfy n players when $n = 3$ and 4 was shown in Section 2.3 and 2.6. This raises the question of whether 5 cuts might also suffice for 5 players. We now show that the answer is no.

Theorem 9.1. *There is no finite algorithm that accomplishes fair division for 5 players using only 5 cuts.*

Proof: If such an algorithm existed, some player, say P_1, would be instructed to make a first cut, say $X = A \cup B$. The algorithm may also specify P_1's evaluation of A and B, but regardless what those values might be we can certainly assume $\mu_1(A) \leq 1/2$ and $\mu_1(B) \leq 1$. In fact, not knowing more about the algorithm, those are the best upper bounds we can give for $\mu_1(A)$ and $\mu_1(B)$. In the tables that follow, numbers listed are upper bounds for the actual values. In fact, in some cases (such as the $2/3$ in the first row of the first table) the number listed is even greater than what we could assume the upper bound to be. This is done for the purpose of reducing the numbers of cases that have to be considered. But, if we show that no algorithm can work even when upper bounds rather than the actual values are placed on the pieces, then the impossibility will be established.

Let us then assume P_1 has made the first cut and upper bounds for the evaluations are given by the table:

	A	B
P_1	$2/3$	1
$P_2, \cdots, P_5$	$2/3$	$1/3$

In this case the values in the second row may occur, since P_1 cut. The algorithm must cover this case.

Since we are allowed only 5 cuts, some 4 players receive a share consisting of a single piece. Consider these 4 players and where their single pieces come from.

Case 1: At least 2 players receive a single subpiece of B.

So B must be cut at least one more time. Look at the first time B is cut. If P_1 cuts B into $B_1 \cup B_2$, then suppose we have the upper bounds:

	A	B_1	B_2
P_1	$2/3$	$1/2$	1
$P_2, \cdots, P_5$	$2/3$	$1/6$	$1/6$

But then none of the players $P_2, \cdots, P_5$ can be satisfied with either B_1 or B_2 or any single subpiece. In the case under consideration, one of $P_2, \cdots, P_5$ will not receive the reguired $1/5$.

So we can assume some other player, say P_2, cuts B into $B_1 \cup B_2$. Then suppose we get:

	A	B_1	B_2
P_1	$2/3$	$5/6$	$1/6$
P_2	$2/3$	$1/3$	$1/6$
$P_3, \cdots, P_5$	$2/3$	$1/6$	$1/6$

Now no player accepts B_2 or any subpiece of B_2, so 2 players (namely P_1 and P_2) must receive a single subpiece of B_1. By Lemma 8.1 we know the algorithm cannot guarantee them both $1/5$, since we must guarantee P_2 a piece of size $1/5 > (1/2)(1/3)$.

Case 2: At most 1 player receives a single subpiece of B.

Then at least 3 players receive a single subpiece of A. We will not need to be concerned with what may happen to piece B because we show we have impossible conditions to satisfy on piece A. Consider the first time A is cut into $A_1 \cup A_2$. We may suppose (without loss of generality) that P_1 cuts and suppose we get:

	A_1	A_2
P_1	$2/3$	$2/3$
$P_2, \cdots, P_5$	$1/3$	$1/3$

At least one more cut must be made on A_1 or A_2 and we may assume (without loss of generality) that A_2 is cut into $A_3 \cup A_4$. If P_1 makes this cut, then suppose we get:

	A_1	A_3	A_4
P_1	$2/3$	$2/3$	$2/3$
$P_2, \cdots, P_5$	$1/3$	$1/6$	$1/6$

Then neither A_3 nor A_4 is acceptable to any of the players $P_2, \cdots, P_5$. Hence, at least 2 of them must receive a single subpiece of A_1. Again by Lemma 8.1 not both can be guaranteed $1/5$.

If some other player (without loss of generality P_2) cuts A_2, then suppose we get:

	A_1	A_3	A_4
P_1	$2/3$	$1/2$	$1/6$
P_2	$1/3$	$1/3$	$1/6$
$P_3, \cdots, P_5$	$1/3$	$1/6$	$1/6$

Now A_4 is not acceptable to any of the players. Hence, since three players are to get single pieces from A, at least 2 players receive a single subpiece of A_1 or at least 2 receive a single subpiece of A_3. Lemma 8.1 again shows not both can be guaranteed $1/5$.

9.3 What Can Be Done with *k* Cuts?

Recall that $M(n, k)$ is the most cake we can guarantee each of the n players when we are allowed only k cuts. From Section 9.2 and earlier algorithms we know $M(2, 1) = 1/2$, $M(3, 3) = 1/3$, $M(4, 4) = 1/4$, and $M(5, 6) = 1/5$. Of course, once we know $M(n, k_0) = 1/n$, then if $k > k_0$, $M(n, k) = 1/n$ since we can't guarantee all n players more than $1/n$ – for example, all of the measures might be the same.

We also know $M(3, 2) < 1/3$, $M(4, 3) < 1/4$, and $M(5, 5) < 1/5$. Can we find exact values, or at least lower bounds, for these values? What can be said about the general cases $M(n, n-1)$, $M(n, n)$, etc.? Although we don't yet know of any general formula for $M(n, k)$, we do know quite a bit about the cases mentioned above. This is summarized in the following theorem.

Theorem 9.2. *If $M(n,k)$ is the most cake that can be guaranteed each of n players using k cuts, then:*

(1) $M(n,n-1) = 1/(2n-2)$ *for* $n \geq 2$,

(2) $M(3,3) = 1/3$ *and*

$M(n,n) = 1/(2n-4)$ *for* $n \geq 4$,

(3) $M(n,n+1) \geq 1/(2n-5)$ *for* $n \geq 5$.

In each case we must first give an algorithm that produces n shares of the appropriate size. These algorithms, showing that portions of the prescribed size can be generated, give only a lower bound for $M(n,k)$. In order to show the equality claimed in Cases (1) and (2), we must also show that larger shares cannot always be guaranteed. The case when $n-1$ cuts are allowed is simpler since we know that every player's share consists of exactly one piece. When n cuts are allowed, the fact that one player will get a share consisting of two pieces complicates matters somewhat. As the number of cuts increases beyond n, the complications of accounting for several players receiving multiple pieces seem to make an exact determination of $M(n,k)$ quite difficult.

All three algorithms are recursive and proved by induction. The case $n=2$ for (1) is accomplished by ordinary Cut and Choose, which gives two players each at least 1/2 with one cut. The cases $n=3$ and $n=4$ for (2) are accomplished by the two algorithms discussed in Sections 2.3 and 2.6. The algorithm in the previous section, which accomplishes simple fair division for 5 players using 6 cuts, establishes (3) for the case $n=5$.

For larger values of n, all three algorithms begin with one player cutting the cake into two equal pieces. Subsets of the players then share one or the other piece. The main differences between the algorithms are the criteria for deciding how to group the players into two subsets.

Since we are using induction, we will be applying the induction hypothesis on pieces where there may be disagreement about size. Nevertheless, we care only about what fraction of the piece a particular player receives. Thus it suffices to normalize the measure and take $\mu_i(A) = 1$ for all i.

Because of the similarities in the proofs, we will give a complete description of the algorithm only for (1).

Proof that $M(n,n-1) \geq 1/(2n-2)$: As was already noted above, Cut and Choose suffices when $n=2$. The proof of the case $n=3$ begins to illustrate the general method but is best handled separately, since it violates one general inequality we will want to use for other values of n.

When $n=3$, have P_3 cut X into pieces $A \cup B$ so that $\mu_3(A) = \mu_3(B) = 1/2$. Then either:

Case (i): Both $\mu_1(A) \leq 1/2$ and $\mu_2(A) \leq 1/2$ (or in the symmetric case $\mu_1(B) \leq 1/2$ and $\mu_2(B) \leq 1/2$); or

Case (ii): If P_1 and P_2 disagree on which piece is smaller, we may assume $\mu_1(A) > 1/2$ while $\mu_2(A) \leq 1/2$.

In Case (i) let P_3 have A, and let P_1 and P_2 share B with one more cut. Both P_1 and P_2 are guaranteed at least 1/4.

In Case (ii) let P_1 have A, and let P_2 and P_3 share B. The reasoning is similar to (i).

Now suppose $n \geq 4$ and we want to guarantee every player at least $1/(2n-2)$. Again begin by P_n cutting $A \cup B$ so that $\mu_n(A) = \mu_n(B) = 1/2$. We may assume without loss of generality that at least $(n-1)/2$ of the remaining players think $\mu(B) \leq 1/2$.

Considering just the $n-1$ non-cutters, we can order the players by their opinions of piece B and assume $\mu_1(B) \leq \mu_2(B) \leq \cdots \leq \mu_{n-1}(B)$. Ask the players in order the following questions until we get the first "yes" answer:

(i) Is $\mu_{n-1}(B) \leq 2/(2n-2)$?

(By the ordering of the players, if the answer is "yes," this means all $n-1$ non-cutters believe B is worth at most $2/(2n-2)$.)

(ii) Is $\mu_{n-2}(B) \leq 4/(2n-2)$?

$\vdots$

(j) Is $\mu_{n-j}(B) \leq 2j/(2n-2)$?

Suppose the answer to question (j) is "no" for some $j > (n-1)/2$. Then more than $(n-1)/2$ players, namely $P_{n-j}, \cdots, P_{n-1}$, think B is worth more than $2j/(2n-2) > 1/2$. This contradicts our earlier assumption. Thus, the first "yes" must occur for some $j \leq (n+1)/2$.

If the answer to question (i) is "yes," then P_n gets piece B while $P_1, \cdots, P_{n-1}$ share piece A using $n-2$ cuts. Since $\mu_i(A) \geq (2n-4)/(2n-2)$ for all i with $1 \leq i \leq n-1$, by the induction hypothesis each is guaranteed at least $(1/(2(n-1)-2)) \times ((2n-4)/(2n-2)) = 1/(2n-2)$.

Suppose the answer to question $(j-1)$ is "no" and to question (j) is "yes," where $2 \leq j \leq (n+1)/2$. Then $\mu_{n-j+1}(B) > 2(j-1)/(2n-2)$ and $\mu_{n-j}(B) \leq 2j/(2n-2)$. Let $P_1, \cdots, P_{n-j}$ share A using $n-j-1$ cuts. By induction each is guaranteed at least $(1/(2(n-j)-2)) \times ((2n-2j-2)/(2n-2)) = 1/(2n-2)$ since $\mu_i(A) \geq (2n-2j-2)/(2n-2)$ for all i with $1 \leq i \leq n-j$. Also $2(n-j)-2 \geq n-3 > 0$, so that $n > 3$ and the induction hypothesis applies.

Also, let $P_{n-j+1}, \cdots, P_{n-1}$ and P_n share B using $j-1$ cuts. By induction each is guaranteed at least $(1/(2j-2)) \times ((2j-2)/(2n-2)) = 1/(2n-2)$, since $\mu_i(B) > (2j-2)/(2n-2)$ for all i with $n-j+1 \leq i \leq n-1$, and $\mu_n(B) = 1/2 \geq (2j-2)/(2n-2)$ since we know $j \leq (n+1)/2$. This completes the algorithm required for (1).

Proof that $M(n, n-1) = 1/(2n-2)$: In light of the algorithm just presented, we must show that *it is impossible to guarantee each of n players strictly more than* $1/(2n-2)$ *of the cake using* $n-1$ *cuts.* Again, the proof is by induction. The result certainly holds for $n = 2$ because the two players cannot both be guaranteed more than $1/2$ of the cake (regardless how many cuts are used), since they may both be using the same measure.

Assuming the italicized statement for values up to $n-1$, suppose $P_1, \cdots, P_n$ are to divide the cake. The argument is made simple because we know each player will receive a portion of cake consisting of a single piece. Someone, say P_n, must make the first cut, $X = A \cup B$, and we may assume $\mu_n(A) \leq 1/2$ regardless of what instructions the algorithm gives. One branch of the algorithm is generated by evaluations $\mu_i(A) = (n-2)/(n-1)$ and $\mu_i(B) = 1/(n-1)$ for all i with $1 \leq i \leq n-1$. Since these players did not cut, those evaluations are possible and the algorithm must complete the division with those values on A and B.

By Lemma 8.1, it is impossible for two or more players to share B, since at least one of them places value $1/(n-1)$ on B. Hence, one player must get all of B and the other $n-1$ players share A using $n-2$ cuts. Regardless of whether P_n is one of these $n-1$ players, all view A as worth at most $(n-2)/(n-1)$. By the induction hypothesis, the most that can be guaranteed to each is $(1/(2(n-1)-2)) \times ((n-2)/(n-1)) = 1/(2n-2)$.

Proof that $M(n, n) \geq 1/(2n-4)$: For Part (2) where we are restricted to n cuts, the algorithm must produce shares of size at least $1/(2n-4)$. That $M(3,3) = 1/3$ and $M(4,4) = 1/4$ has been established in Sections 2.3 and 2.6 respectively. We also want to treat the cases $n = 5$ and 6 separately. They require only minor modifications of the general procedure given below, and are left for Exercises 9.1 and 9.2.

Assuming the cases for $n = 4, 5$, and 6 have been established, suppose we know the result holds for k players with $4 \leq k \leq n-1$. Given n players, $n \geq 7$, have P_n cut A and B both of size 1/2. Just as in the algorithm that used only $n-1$ cuts, we may assume at least $(n-1)/2$ of the non-cutters think $\mu(B) \leq 1/2$ and $\mu_1(B) \leq \mu_2(B) \leq \cdots \leq \mu_{n-1}(B)$. Again there must be some smallest j, $1 \leq j \leq (n+1)/2$, such that

(i) $\mu_{n-j+1}(B) > (2j-2)/(2n-4)$ and

(ii) $\mu_{n-j}(B) \leq 2j/(2n-4)$.

As in the previous algorithm, if (ii) is false for $j > (n-1)/2$, then more than $(n-1)/2$ non-cutters think $\mu(B) > 2j/(2n-4) > (n-1)/(2n-4) > 1/2$, a contradiction. Also, if (i) and (ii) hold for $j = 1$, then all non-cutters believe $\mu(B) \leq 2/(2n-4)$.

Case 1: $1 \le j \le (n-1)/2$.

Let $P_1, \cdots, P_{n-j}$ share A using $n-j$ cuts. They each think $\mu(B) \le 2j/(2n-4)$, so $\mu(A) \ge (2n-2j-4)/(2n-4)$ and each gets at least $1/(2(n-j)-4) \times (2n-2j-4)/(2n-4) = 1/(2n-4)$. Since $n \ge 7, n-j \ge (n+1)/2 \ge 4$ so the induction hypothesis applies.

Now using the algorithm for (1) above, let $P_{n-j+1}, \cdots, P_n$ share B using $j-1$ cuts. They all think $\mu(B) \ge (2j-2)/(2n-4)$, so each gets at least $1/(2j-2) \times (2j-2)/(2n-4) = 1/(2n-4)$.

Since $\mu_n(B) = 1/2 > (2j-2)/(2n-4)$ for $j < (n-1)/2$, player P_n can be included in this group. If $j = 1$, P_n is the only player in this group and is clearly satisfied.

Case 2: $n/2 \le j \le (n+1)/2$.

Let $P_1, \cdots, P_{n-j}, P_n$ share A using $n-j+1$ cuts. If $\mu_{n-j}(B) > 1/2$, then only non-cutters $P_1, P_2, \cdots, P_{n-j-1}$ can think $\mu_j(B) \le 1/2$. But $n-j-1 \le n-(n/2)-1 < (n-1)/2$ in which case fewer than $(n-1)/2$ of the non-cutters think $\mu(B) \le 1/2$, a contradiction. Thus all $n-j+1$ players in this group think $\mu(B) \le 1/2$. Each gets at least $(1/(2(n-j+1)-4)) \times (1/2) = 1/(4n-4j-4) \ge 1/(2n-4)$. Since $n \ge 7$, $n-j+1 \ge n+1-(n+1)/2 = (n+1)/2 \ge 4$ so the induction hypothesis applies. Using the algorithm for (1), let $P_{n-j+1}, \cdots, P_{n-1}$ share B using $j-2$ cuts. Each gets at least $(1/(2(j-1)-2)) \times ((2j-2)/(2n-4)) > 1/(2n-4)$.

The description of the algorithm for (3) proceeds along similar lines as the two above for (1) and (2).

Proof that $M(n, n) = 1/(2n-4)$: An extension of Lemma 8.1 is required.

Lemma 9.1. *There is no finite algorithm for n players, $n \ge 3$, using $n-1$ cuts that guarantees portions of size greater than $1/(2n-4)$ to $n-1$ of the players while giving the other player a portion with positive measure.*

Proof of Lemma 9.1. For $n = 3$ the result is clear, since all three may use the same measure. Proceed by induction and assume the result for $n-1$ players where $n \ge 4$. We may assume without loss of generality that the algorithm for n players directs P_n to make the first cut, $X = A \cup B$, and that $\mu_n(A) \le 1/2$. Suppose $\mu_i(A) = (n-3)/(n-2)$ and $\mu_i(B) = 1/(n-2)$ for i with $1 \le i \le n-1$. By Lemma 8.1 two or more non-cutters cannot be in the group of players who take their portions from B. Also applying the induction hypothesis, we see that it is impossible for $n-1$ players to share A with $n-2$ of them getting more than $1/(2(n-1)-4) \times (n-3)/(n-2) = 1/(2n-4)$ while the other gets a piece of positive measure. (Note for P_n, $1/2 \leq (n-3)/(n-2)$.)

This result only leaves the possibility that B is shared by P_n, who may get more than $1/(2n-4)$, and some non-cutter who receives a piece with positive measure. But this requires the other $n-2$ non-cutters to share A using $n-3$ cuts and each receive more than $1/(2n-4) = 1/(2(n-2)-2) \times (n-3)/(n-2)$, which is impossible from (1) of the theorem.

That $M(n,n) \leq 1/(2n-4)$ for $n = 3$ and 4 is trivial. So assume the inequality for $n-1$ or fewer players with $n \geq 5$. As above, assume the algorithm directs P_n to cut, $X = A \cup B$, with $\mu_n(A) \leq 1/2$ while $\mu_i(A) = (n-3)/(n-2)$ and $\mu_i(B) = 1/(n-2)$ for i with $1 \leq i \leq n-1$.

Case 1: All players receive their portions entirely from A or entirely from B.

If one player gets all of B, then the other $n-1$ players are required to share A using $n-1$ cuts, in which case, by the induction hypothesis, the most each can be guaranteed is $1/(2(n-1)-4) \times (n-3)/(n-2) = 1/(2n-4)$. (Again $1/2 \leq (n-3)/(n-2)$.)

Also, two or more non-cutters cannot be in the group sharing B since they may use equal measures. The only other possibility under Case 1 is for P_n and some non-cutter, say P_{n-1}, to share B. They can't both get more than $1/(2n-4)$ with one cut. If P_n cuts, P_{n-1} may view the two pieces as each worth $1/(2n-4)$. If P_{n-1} cuts a piece larger than $1/(2n-4)$, P_n may view the complementary piece having value less than $1/(2n-4)$. On the other hand, if two cuts are used on B, then the remaining $n-2$ non-cutters must share A using only $n-3$ cuts. From (1), the most they can receive is $1/(2(n-2)-2) \times (n-3)/(n-2) = 1/(2n-4)$.

Case 2: Some player P_j gets a portion consisting of one piece from A and one piece from B.

First note that P_j can't get all of B, otherwise $n-1$ cuts are used to divide A, with P_j receiving one piece from A. Suppose this is accomplished so that each player other than P_j receives more than $1/(2n-4)$. Alter this $n-1$ cut algorithm on A by simply picking some player P_i, $i \neq j$, and awarding him what he already received along with the portion of A originally given to P_j. We now have an $n-1$ cut, $n-1$ players, algorithm on A giving all $n-1$ players (other than P_j) more than $1/(2n-4) = 1/(2(n-1)-4) \times (n-3)/(n-2)$, thereby violating the induction hypothesis.

As before, we can't have two or more non-cutters get their one piece shares from B. On the other hand, if P_n and some non-cutter P_i, $i \neq j$, get their one piece portions from B, then P_i must make the first cut on B. If someone else cuts first, P_i may judge the two resulting pieces to each have value $1/(2n-4)$. So P_i must cut a portion from B that he or she values greater than $1/(2n-4)$ and P_i will not accept even all of the complementary piece as having value $1/(2n-4)$. But all other players sharing B with P_i may also value that complementary piece at less than $1/(2n-4)$.

The only case left is for some single player P_i, $i \neq j$, to get his or her one piece share from B. But that leaves P_j and the other $n-2$ players to share A using $n-2$ cuts. By Lemma 9.1, P_j cannot receive a portion with positive measure while the others receive more than $1/(2(n-1)-4)\times(n-3)/(n-2) = 1/(2n-4)$. This requires both P_i and P_j to have received portions from B which they each consider as worth more than $1/(2n-4)$. We have already observed that this is impossible. This completes the proof of Theorem 9.2.

Combining information from Section 8.2 and Theorem 9.2, some things can be said about $N(n,k)$, the number of players who can be guaranteed $1/n$ of the cake when k cuts are used. In particular, when there are $2n-2$ players, the goal at hand is to give as many as possible a piece of size $1/(2n-2)$. If $n-2$ of them are given nothing, each of the remaining n can be given $1/(2n-2)$ of the cake with $n-1$ cuts using the algorithm of Theorem 9.2 (1), which is certainly as many as possible using $n-1$ cuts, since there are only n pieces to give away. Therefore we have:

Corollary 9.1. *For* $n \geq 2, N(2n-2, n-1) = n$.

So it follows that $N(4,2) = 3, N(6,3) = 4, N(8,4) = 5$, etc., and from Theorem 8.2 we have $N(4,3) = 3$ (so having three cuts available instead of two does not increase the number of players who can be given $1/4$ of the cake). In general $N(n, n-1) < n$ for $n \geq 3$.

With these sketchy beginnings, much remains to be discovered about $N(n,k)$.

9.4 Divide and Conquer Revisited

Pursuing Steinhaus' statement "Interesting mathematical problems arise if we are to determine the minimal number of cuts necessary for (simple) fair division" when using a finite algorithm, let us define $F(n)$ to be that minimal number of cuts for simple fair division for n players. Only the first few values for F are known, namely, $F(1) = 0$, $F(2) = 1$, $F(3) = 3$, $F(4) = 4$, and $F(5) = 6$. (See the table in Section 7.2.)

Moreover, in Section 2.4 the Divide and Conquer Algorithm was given which accomplished simple fair division for n players using $D(n)$ cuts where $D(n) =$

$nk-2^k+1$, $k = \lceil \log_2 n \rceil$, with initial value $D(1) = 0$. (See Exercise 2.6.) Since $F(n) \leq D(n)$, $F(n) = O(n \log_2 n)$. Since we know $F(4) = 4 < D(4) = 5$, the question of improving the bound given by $D(n)$ arises even for small numbers of players.

Recall that the Divide and Conquer Algorithm for n players used $n-1$ cuts to separate the n players into two groups of nearest possible equal size so that each group could independently divide complementary portions of the cake. This led to the recursion

$$D(n) = (n-1) + D\left(\lfloor (n-1)/2 \rfloor\right) + D\left(\lceil (n-1)/2 \rceil\right),$$

with $D(1) = 0$.

A simple modification of the Divide and Conquer Algorithm presents possibilities for improving the bounds for $F(n)$. Rather than cut near halves (in ratio $k : k$ when $n = 2k$ or $k : k+1$ when $n = 2k+1$) and separate the players into two (nearly) equal size groups to independently divide their portions, we can instruct the first cutter to divide in the integer ratio $t : n-t$. Then, using $n-1$ cuts, two groups of size t and $n-t$ can be formed that independently divide the two resulting complementary portions. This leads to the modified recursion

$$F(n) \leq (n-1) + F(t) + F(n-t).$$

It can be shown (see Exercise 9.3) that if the values for F are convex for up to $n-1$ players, then the optimal division from the recursion for n players is produced by a near-halves cut, i.e., choose $t = \lfloor n/2 \rfloor$. (Also see Exercise 2.2.) But the values $F(1) = 0$, $F(2) = 1$, $F(3) = 3$, and $F(4) = 4$ are not convex, and in fact smaller bounds in some cases are produced by other values of t. A table of the first few values illustrates the point.

Small Case Bounds for $F(n)$																
n	1	2	3	4	5	6	7	8	9	10	11	12	13	14	15	16
upper bound for $F(n)$	0	1	3	4	6	8	13	15	18	21	24	27	33	36	40	44

We have observed that through $n = 5$ the bounds are exact. For $n = 4, 5$, and 6 we have seen *ad hoc* algorithms giving improvements on the number of cuts produced by the Divide and Conquer Algorithm. The entry for $n = 7$ is from recursion and can no doubt be improved.

All bounds for n with $7 \leq n \leq 13$ are produced for t chosen near halves. For example, $F(12) \leq 11 + F(6) + F(6) \leq 27$ and $F(13) \leq 12 + F(6) + F(7) = 33$. Yet $F(14) \leq 13 + 2F(7) \leq 39$ is the bound produced for $t = 14/2$, while the better bound $F(14) \leq 13 + F(6) + F(8) = 36$ is generated by choosing $t = 6$. The bounds given for $n = 15$ and 16 are also produced for values of t other than $\lfloor n/2 \rfloor$.

The Divide and Conquer Recursion generates the dominating $n \log_2 n$ term in the bound for $F(n)$ and a new approach will be required to improve that. Nevertheless, the lower order terms do offer some possibility of tightening the bounds on $F(n)$. To illustrate this we can show that

$$F(n) \leq n \log_2 n - 1.12n$$

for $n \geq 8$. It can be routinely checked that the inequality holds for n with $8 \leq n \leq 15$. By induction we can prove that if $F(n) \leq n \log_2 n - cn$ for fixed c and n satisfying $t \leq n \leq 2t-1$, then in fact the inequality holds for all n with $t \leq n$. The justification requires only some straightforward algebra as follows:

If $n = 2k$, then

$$\begin{aligned} F(n) &= F(2k) \leq (2k-1) + 2F(k) \leq (2k-1) + 2k\log_2 k - 2ck \\ &= 2k \log_2 2k - 2ck - 1 < n \log_2 n - cn. \end{aligned}$$

If $n = 2k+1$ then

$$\begin{aligned} F(n) = F(2k+1) &\leq 2k + F(k) + F(k+1) \\ &\leq 2k + k\log_2 k + (k+1)\log_2(k+1) - ck - c(k+1) \\ &\leq 2k + k\log_2(k+1/2)^2 + \log_2(k+1) - c(2k+1) \\ &= (2k+1)\log_2(2k+1) - c(2k+1) \\ &\quad - \log_2((2k+1)/(k+1)) \\ &\leq n\log_2 n - cn. \end{aligned}$$

Improvements in bounds for $F(n)$ with n small can produce improvements in the value $c = 1.12$.

Still, the over-all state of the problem now is that $F(n) = O(n \log_2 n)$ is guaranteed by divide and conquer techniques, and any significant improvement on this bound will require completely new insights about simple fair division.

9.5 EXERCISES

9.1. Show that five players can be guaranteed a share of at least $1/6$ of the cake using only 5 cuts.

9.2. Show that six players can be guaranteed a share of at least $1/8$ of the cake using only 6 cuts.

9.3. Prove that the values generated by the recursion $D(n) = (n-1) + D(\lfloor n/2 \rfloor) + D(\lceil n/2 \rceil)$ with $D(1) = 0$ and $D(2) = 1$ are convex.

9.4. Prove that if the values for $G(t)$ are convex for $t = 1, 2, \cdots, n-1$ and $G(n) \leq (n-1) + G(t) + G(n-t)$ for all t with $1 \leq t \leq n-1$, then the best bound for $G(n)$ is given when $t = \lfloor n/2 \rfloor$.

9.6 PROJECTS

9.1. As far as we know, only superficial attention has been given to the value $N(n, k)$. There may be some reasonably accessible results that would be of interest.

9.2. Resolve the discrepancy between the facts that there is some meager evidence that $\mathrm{O}(n)$ cuts suffice for fair division, yet the most efficient algorithm known uses $\mathrm{O}(n \log_2 n)$ cuts.

TEN

Exact and Envy-Free Algorithms

10.1 Resetting the Stage

In Section 8.3, it was established that no finite algorithm can accomplish *exact* division, even for two players and equal shares. In this chapter, we will give a finite algorithm that comes as close as we wish to exact division for any number of players and in any ratio. This extends what was accomplished on near-exact division in Section 5.4. Furthermore, once the desired tolerance of error is set, the algorithm is bounded with the number of steps used a function of the size of that acceptable tolerance.

The proof of this result uses a very nice and intuitive geometric result. If we have a number of short vectors in Euclidean space E^d whose sum is the zero vector, it is possible to order the vectors so that all partial sums of the vectors, taken in that order, do not stray far from the origin.

The Near-Exact Algorithm will then be used to produce an unbounded, finite algorithm for envy-free division for any number of players and in any ratios of shares. After that is accomplished we present a second finite, unbounded algorithm for envy-free division given by Brams and Taylor. Recall that *bounded finite* algorithms for envy-free division are known only for 2 or 3 players (Section 1.4), so the question of bounded finite algorithms for more than 3 players is still an open question.

The chapter concludes with work due to Julius Barbanel giving necessary and sufficient conditions that super envy-free portions *exist*, i.e., portions for which each player feels every other player is slighted. Finally, given portions satisfying a certain disagreement condition, an algorithm is described that actually produces such portions.

10.2 Near-Exact Division

In Section 5.4, a Near-Exact Algorithm was described for two players in any ratio of shares. Now we extend the result to any number of players. The idea of the proof is unchanged, and essentially all that is needed is to extend the methods used on short vectors in the plane to d-dimensional space E^d.

The theorem needed was published in 1930 by Bergstrom [Ber]. He showed that if you have a zero sum set of vectors in E^d each of which has magnitude bounded by M, there is a constant $K(d)$, depending only on the dimension d, so that if the vectors are properly ordered, $v_1, v_2, \cdots, v_t$, then $\|\sum_{i=1}^{k} v_i\| \leq M \cdot K(d)$ for all $k \leq t$. He gives bounds $K(2) = \sqrt{5}/2$ and in general $K(d) = \mathrm{O}\ (\sqrt{d})$. In [KK] it is shown that one can take $K(d) = d$. Forgoing the effort needed to establish these stronger bounds, we present a simpler proof using Bergstrom's essential geometric ideas to show $K(d)$ can be chosen to be 3^{d-1}.

Theorem 10.1. *Given a set of vectors $V = \{v_i\}_{i=1}^{t}$ in E^d such that for all i, $\|v_i\| \leq M$, and $\sum_{i=1}^{t} v_i = 0$, there is a permutation Π of $\{1, 2, \ldots, t\}$ such that all of the vectors $w_r = \sum_{i=1}^{r} v_{\Pi(i)}$ have magnitude $\|w_r\| \leq M \cdot 3^{d-1}$.*

Proof: The proof is by induction on d. For $d = 1$, start with any vector and choose a next vector to be negative if the current partial sum is positive and vice versa. If the vectors lie in E^d, for each point θ on the unit sphere S^{d-1} in E^d, let $L(\theta)$ be the line through θ and the origin and $H(\theta)$ the hyperplane passing through the origin which is perpendicular to $L(\theta)$. Let $H_1(\theta)$ and $H_2(\theta)$ be the corresponding open half spaces. Partition V into the three disjoint subsets $V_0(\theta) = \{v \in V : v \in H(\theta)\}, V_i(\theta) = \{v \in V : v \in H_i(\theta)\}, i = 1, 2$, and let $y_i(\theta) = \sum_{v \in V_i(\theta)} v,\ i = 0, 1, 2.$

If $z_i(\theta), i = 1, 2$, is the projection of y_i on $L(\theta)$, we see that z_i is a continuous function of θ and $z_1 = -z_2$. Choose z_i^* where $|z_i^*| = \sup_{\theta \in S^{d-1}} |z_i|$; let θ^* be the corresponding direction and y_i^* the corresponding vector in $H_i(\theta^*)$ whose projection on $L(\theta^*)$ is z_i^*.

We first show $y_i^* = z_i^*$. Otherwise, let $\theta' \in S^{d-1}$ be in the direction of y_i^*. Then any vector in $V_1(\theta')\backslash V_1(\theta^*)$ has positive component in the direction of y_1^*, while any vector in $V_1(\theta^*)\backslash V_1(\theta')$ has non-positive component in the direction of y_1^*. Thus $|z_1(\theta')| > |z_1(\theta^*)|$ which is a contradiction.

If x_i is the projection of v_i onto $H(\theta^*)$, then $\sum_{v_i \in H_1(\theta^*)} x_i = \sum_{v_i \in H_2(\theta^*)} x_i = 0$.

Also $\sum_{v_i \in H(\theta^*)} v_i = 0$. By induction each of the set of vectors $\{x_i : v_i \in H_1(\theta^*)\} \cup \{v_i : v_i \in H(\theta^*)\}$ and $\{x_i : v_i \in H_2(\theta^*)\}$ can be rearranged so that neither list produces a partial sum farther than $M(3^{d-2})$ from the origin. Denote the set of vectors v_i used in the first union as U_1 and $V - U_1 = U_2$, and assume both sets have been so ordered.

The desired rearrangement for V can now be described. Take vectors in order from U_1 until $H_1(\theta^*)$ is first entered; next add vectors in order from U_2 until $H_2(\theta^*)$ is first entered; then add vectors in order from the remaining vectors in U_1 until $H_1(\theta^*)$ is first re-entered; repeat until the vectors are all chosen. The resulting partial sums are bounded by $\sqrt{M^2 + 2^2M^2(3^{d-2})^2} < M \times 3^{d-1}$ and the result is established.

We are now ready to describe the algorithm for near-exact division. First some definitions are required.

A cake division among n players is *exact in the ratios* $\alpha_1 : \alpha_2 : \cdots : \alpha_n$ provided $\mu_i(X_j)/\mu_i(X) = \alpha_j/(\alpha_1 + \cdots + \alpha_n)$, whenever $1 \le i, j \le n$.

A cake division among n players is ϵ *near-exact in the ratios* $\alpha_1 : \alpha_2 : \cdots : \alpha_n$ provided $|\mu_i(X_j)/\mu_i(X) - \alpha_j/(\alpha_1 + \cdots + \alpha_n)| < \epsilon$, whenever $1 \le i, j \le n$.

We now give an algorithm for ϵ near-exact division among n players. The procedure is exactly the same as that given in Section 5.4 for two players. We will first cut pieces that all players agree are small. Then, using Theorem 10.1, the pieces will be arranged in an order $X_1, X_2, X_3, \cdots, X_t$ so that for each j the values $\mu_i(X_1 \cup \cdots \cup X_j)$ are nearly equal for all i. This will allow the list of t small pieces to be partitioned into n sublists having the required values. We note that the number of steps used is bounded as a function of ϵ.

Algorithm for ϵ Near-Exact Division for n Players and Any Ratios [RW4]

Theorem 10.2. *Let $\mu_1, \mu_2, \ldots, \mu_n$ be probability measures on a cake X. Given $\epsilon > 0$ and positive numbers $\alpha_1, \ldots, \alpha_n$ with $\sum \alpha_i = 1$, there is a bounded finite algorithm producing a partition $X = X_1 \cup X_2 \cup \ldots \cup X_n$ that is ϵ near-exact in the ratios $\alpha_1 : \alpha_2 : \cdots : \alpha_n$.*

Proof: Assume $\mu_1(X) = \ldots = \mu_n(X) = 1$. First have P_1 cut X into k pieces with equal μ_1 measure, where k will be chosen later, and then have P_2 reduce (by no more than $k-1$ cuts) those pieces so that no piece has a μ_2 measure exceeding $1/k$. Repeat for all n players to produce the partition $X = A_1 \cup A_2 \cup \ldots \cup A_t$, $t \leq nk$, with $\mu_i(A_j) \leq 1/k$ for all i, j. Set $x_j = \frac{1}{n} \sum_{i=1}^{n} \frac{\mu_i(A_j)}{\mu_i(X)}$ and define ϵ_{ij} by $\frac{\mu_i(A_j)}{\mu_i(X)} = x_j + \epsilon_{ij}$. We see that $\sum_{j=1}^{t} x_j = \frac{1}{n} \sum_{j=1}^{t} \sum_{i=1}^{n} \frac{\mu_i(A_j)}{\mu_i(X)} = \frac{1}{n} \sum_{i=1}^{n} \sum_{j=1}^{t} \frac{\mu_i(A_j)}{\mu_i(X)} = 1$ and $\sum_{j=1}^{t} \epsilon_{ij} = \sum_{j=1}^{t} \left(\frac{\mu_i(A_j)}{\mu_i(X)} - x_j \right) = 0$. Also, $|\epsilon_{ij}| \leq \frac{1}{k}$ since both x_j and $\mu_i(A_j)$ are bounded above by $1/k$.

Then whenever $1 \leq j \leq t$, the vector $v_j = (\epsilon_{1j}, \ldots, \epsilon_{nj})$ satisfies $\|v_j\| \leq \frac{\sqrt{n}}{k}$. Assume the vectors v_j are ordered (which can be done using a number of steps depending on nk) so that Lemma 10.1 applies (using $k(d) = d$ as given in [KK] for simplicity). Then $\|\sum_{j=0}^{r} v_j\| \leq \frac{n^{3/2}}{k}$ whenever $1 \leq r \leq t$. Since for all p, $0 \leq x_p \leq \frac{1}{k}$, there is an increasing sequence $0 = t_0 < t_1 < t_2 < \ldots < t_n = t$ so that $|\sum_{p=t_{i-1}+1}^{t_i} x_p - \alpha_i| < \frac{1}{k}$, for $1 \leq i \leq n$. Set $X_j = \bigcup_{p=t_{j-1}+1}^{t_j} A_p$. Then $|\mu_i(X_j) - \alpha_j| = |\sum_{p=t_{j-1}+1}^{t_j} (x_p + \epsilon_{ip}) - \alpha_j| \leq \frac{1}{k} + \frac{2n^{3/2}}{k} = \frac{1}{k}(1 + 2n^{3/2})$. The proof is complete by choosing $k > \frac{1}{\epsilon}(1 + 2n^{3/2})$.

10.3 Envy-Free Division

Using the Near-Exact Algorithm just described, we first give a finite, unbounded algorithm for envy-free division for n players in any ratios. The procedure is described below.

One player will cut the cake in the indicated ratios. If all n players agree on all the cuts, fine! Assign the pieces and we are done. (Agreement is sometimes serendipitous too.) More than likely there is disagreement somewhere; isolate a single piece A where disagreement occurs and split the players into two camps, those who think A is larger and those who think it is smaller.

Next, by a near-exact (among all n players) division, supplement A with another portion of the cake B so that those who judged A larger have (barely) sufficient cake in $A \cup B$ to more than cover their combined shares, while the ones who judged A smaller have (slightly) more than enough cake in the complementary set $X - (A \cup B)$ to cover their combined shares.

At this point we have the two camps each collectively claiming complementary portions of the cake and each camp is more than satisfied with their collective portion. It remains to divide the cake in each of the two camps, and this is done envy-free (among the players in each camp) and near-exact (among all players) by induction. The resulting over-all division is both near-exact and envy-free. This completes the outline of what is to be done. Unfortunately, the necessary details are not so pleasant [RW4].

A Finite Unbounded Algorithm for ϵ Near-Exact and Envy-Free Division for n Players in Any Ratios

The proof is by induction on the number of players, which introduces a complicating factor. Indeed, as a piece is being shared in an envy-free manner by a subset of the players, the remaining players who don't get a share of this piece must remain satisfied with the over-all scheme. This problem is addressed by proving an even stronger result that allows for observers who don't get any cake but who must view the division as near-exact. Although we may assume that the measures are normalized so that $\mu_i(X) = 1$ on the whole cake, we will not do so for the piece X in the theorem below in order to be able to apply the induction hypothesis to smaller pieces. We may assume, however, that X and consequently all subpieces of X have measure at most one.

Theorem 10.3. *Given any piece of cake X, any $\epsilon > 0$, players $P_1, \cdots, P_n, Q_1, \cdots, Q_m$, and any ratio of positive numbers $\alpha_1 : \alpha_2 : \cdots : \alpha_n$, there is a finite algorithm assigning pieces $X_1, \cdots, X_n$ to the players $P_1, \cdots, P_n$ respectively that is:*

(1) Envy-free among $P_1, \cdots, P_n$, and

(2) ϵ near-exact among $P_1, \cdots, P_n$ and $Q_1, \cdots, Q_m$.

Proof: The theorem is trivial for $n = 1$ and any m, since P_1 gets all of X. Assume the theorem is true for all m any time fewer than n players share X.

Using Theorem 10.2 we can produce an ϵ near-exact partition $X_1, \cdots, X_n$ in the ratios $\alpha_1 : \alpha_2 : \cdots : \alpha_n$ where all $n+m$ players $P_1, \cdots, P_n, Q_1, \cdots, Q_m$ view the division as ϵ near-exact. (The proof of Theorem 10.2 is carried out in E^{n+m} but only n shares are cut.) Moreover, by allowing P_1 to make a last series of cuts we may assume $\mu_1(X_i) = \alpha_i \mu_1(X)$ for $1 \leq i \leq n$. If all of the players $P_2, \cdots, P_n$ agree with P_1, that is, if $\mu_j(X_i) = \alpha_i \mu_j(X)$ whenever $1 \leq i \leq n$ and $2 \leq j \leq n$, then the partition of X is also envy-free and we are done.

Otherwise, there is a piece A that has been cut on which there is disagreement among some of $P_1, \cdots, P_n$. But cutting A into smaller pieces if necessary, we may assume that A is such that $\mu(A)/\mu(X) < \epsilon/4$, where μ without a subscript denotes the measure of any of the players $P_1, \cdots, P_n$ or $Q_1, \cdots, Q_m$ throughout the proof. Also, since there is some disagreement, there is a k, $1 \leq k < n$, such that:

$$\frac{\mu_1(A)}{\mu_1(X)} = \cdots = \frac{\mu_k(A)}{\mu_k(X)} = a > b = b_{k+1} = \frac{\mu_{k+1}(A)}{\mu_{k+1}(X)} \geq \cdots \geq \frac{\mu_n(A)}{\mu_n(X)} = b_n.$$

In particular, we may also assume $a/(1-a) < \epsilon/4$.

We then divide the remaining cake $X - A$ into pieces B and C in a near-exact manner, have $P_1, \cdots, P_k$ share $A \cup B$ in an envy-free manner, have $P_{k+1}, \cdots, P_n$ share C in an envy-free manner, and maintain a close enough degree of near-exactness to guarantee envy-freeness among all of $P_1, \cdots, P_n$. This is possible since $P_1, \cdots, P_k$ view A as larger than any of $P_{k+1}, \cdots, P_n$.

To this end, scale the α_i so that $\alpha_1 + \cdots + \alpha_n = 1$, and let $\alpha_1 + \cdots + \alpha_k = \alpha$. If we set $m_1 = \min\{\alpha_i\}$, $m_2 = \min\{\mu(X)\}$, and $\Delta = \min\{\epsilon/4, m_1(a-b)/2\}$, we can choose ϵ_1 so that $0 < \epsilon_1 < \min\{m_2\,\epsilon/8, m_1 m_2 \Delta(1-a)/2\}$.

We will write $x = y + \Omega(\epsilon_1)$ only if $|x - y| < \epsilon_1$. In particular, if $x = y + \Omega(\epsilon_1)$ and $|\beta| \leq 1$ then $\beta x = \beta y + \Omega(\epsilon_1)$.

We now partition $X - A = B \cup C$ in ϵ_1 near-exact ratio $\dfrac{\alpha - a}{1-a} + \Delta : \dfrac{1-\alpha}{1-a} - \Delta$ so we know that $\dfrac{\mu(B)}{\mu(B \cup C)} = \dfrac{\alpha - a}{1-a} + \Delta + \Omega(\epsilon_1)$ and $\dfrac{\mu(C)}{\mu(B \cup C)} = \dfrac{1-\alpha}{1-a} - \Delta + \Omega(\epsilon_1)$.

Finally, by induction $P_1, \cdots, P_k$ can share $A \cup B = X_1 \cup \cdots \cup X_k$ envy-free and ϵ_1 near-exact in the ratio $\alpha_1 : \cdots : \alpha_k$, and $P_{k+1}, \cdots, P_n$ can share $C = X_{k+1} \cup \cdots \cup X_n$ envy-free and ϵ_1 near-exact in the ratio $\alpha_{k+1} : \cdots : \alpha_n$. Thus, $\dfrac{\mu(X_i)}{\mu(A \cup B)} = \dfrac{\alpha_i}{\alpha} + \Omega(\epsilon_1)$ for $1 \leq i \leq k$, and $\dfrac{\mu(X_h)}{\mu(C)} = \dfrac{\alpha_h}{1-\alpha} + \Omega(\epsilon_1)$ for $k < h \leq n$.

It must be shown that no player $P_1, \cdots, P_k$ envies a piece given to any of the players $P_{k+1}, \cdots, P_n$ or vice versa, and that the division is ϵ near-exact.

Whenever $1 \le i, j \le k < h, l \le n$

$$\begin{aligned}
\mu_i(X_j) &= \frac{\alpha_j}{\alpha}\mu_i(A \cup B) + \Omega(\epsilon_1) \\
&= \frac{\alpha_j}{\alpha}[a\mu_i(X) + \mu_i(B \cup C)\left(\frac{\alpha - a}{1-a} + \Delta\right) + \Omega(\epsilon_1)] + \Omega(\epsilon_1) \\
&= \frac{\alpha_j}{\alpha}[a\mu_i(X) + \mu_i(X)(1-a)\left(\frac{\alpha - a}{1-a} + \Delta\right)] + 2\Omega(\epsilon_1)
\end{aligned}$$

and so

$$\mu_i(X_j) - \alpha_j\mu_i(X) = \frac{\alpha_j}{\alpha}\mu_i(X)(1-a)\Delta + 2\Omega(\epsilon_1). \tag{10.1}$$

Similarly,

$$\mu_i(X_h) - \alpha_h\mu_i(X) = -\frac{\alpha_h}{1-\alpha}\mu_i(X)(1-a)\Delta + 2\Omega(\epsilon_1), \tag{10.2}$$

$$\mu_l(X_h) - \alpha_h\mu_l(X) = \alpha_h\mu_l(X)\left[\frac{a - b_l}{1-a} - \Delta\left(\frac{1-b_l}{1-\alpha}\right)\right] + 2\Omega(\epsilon_1), \tag{10.3}$$

$$\mu_l(X_j) - \alpha_j\mu_l(X) = -\frac{\alpha_j}{\alpha}\mu_l(X)\frac{(1-\alpha)(a-b_l)}{1-a} + \frac{\alpha_j}{\alpha}(1-b_l)\mu_l(X)\Delta + 2\Omega(\epsilon_1). \tag{10.4}$$

Also, for any measure μ, including those associated with the Q_i, and any j, $1 \le j \le n$:

$$\mu(X_j) - \alpha_j\mu(X) = (\Omega\left(\frac{a}{1-a}\right) + \Omega(\Delta))\mu(X) + \Omega(\mu(A)) + 2\Omega(\epsilon_1). \tag{10.5}$$

The proof is completed by verifying that, with the choices of a, Δ, and ϵ_1 above, the right-hand sides of (1) and (3) are positive while (2) and (4) are negative, making the division envy-free, and the right-hand side of (5) has magnitude less than $\epsilon\mu(X)$, guaranteeing ϵ near-exactness. We omit the details.

To accomplish envy-free fair division in ratios $\alpha_1 : \ldots : \alpha_n$ among $P_1, \ldots, P_n$ apply the theorem with $m = 0$. Thus we have:

Theorem 10.4. There is a finite algorithm that accomplishes envy-free and ϵ near-exact division among n players in given ratios $\alpha_1 : \alpha_2 : \ldots : \alpha_n$.

Near-exact division is a useful process that can be utilized in a number of settings. We now present another example. The methods given in Section 5.2 and the proof of Theorem 10.2 show how one can capitalize by using a piece on which there is some disagreement to accomplish certain kinds of division. The following result is in that same spirit: if we can find a piece that one player thinks has zero value while a second player disagrees, then starting with that piece there is a *finite* algorithm that will accomplish *exact* division in an arbitrary ratio $\alpha_1 : \alpha_2$. Of course in general, such a piece may not exist and the general case for *exact* division by *finite* algorithms is still covered by the impossibility results of Section 8.3.

The outline of the exact division in the ratio $\alpha_1 : \alpha_2$ is as follows. Assume $\mu_1(A) = 0$ while $\mu_2(A) = a > 0$. Have the two players cut $X - A = Y_1 \cup Y_2$ near-exact in the ratio $\alpha_1 - \epsilon : \alpha_2 + \epsilon$ where ϵ is chosen small with respect to a.

At this stage P_1 will think Y_1 is a bit too small but with value near α_1. Player P_2 will think Y_2 is smaller than the required value α_2, but piece A has sufficient value to cover that deficit if all of A is given to P_2. A near-exact portion $Y_2^{'}$ of Y_2 is then cut which (i) has sufficient value to P_1 so that $\mu_1(Y_1 \cup Y_2^{'}) > \alpha_1$ and (ii) has small enough value to P_2 so that $\mu_2(Y_2 - Y_2^{'}) + a > \alpha_2$ (i.e., the value of A still covers the deficit even after $Y_2^{'}$ is given away).

Next P_1 cuts a piece $Y_2^{''}$ from $Y_2^{'}$ so that a share satisfying $\mu_1(Y_1 \cup Y_2^{''}) = \alpha_1$ results. Player P_1 is now exactly satisfied with $Y_1 \cup Y_2^{''}$ but player P_2 still feels the portion is smaller than α_1. Player P_2 is now able to cut a portion A_1 from A so that $\mu_2((Y_2 - Y_2^{''}) \cup A_1) = \alpha_2$ and therefore $\mu_2(Y_1 \cup Y_2^{''} \cup (A - A_1)) = \alpha_1$. Giving $A - A_1$ to P_1 does not change the value $\mu_1(Y_1 \cup Y_2^{''}) = \mu_1(Y_1 \cup Y_2 \cup (A - A_1)) = \alpha_1$ and exact division results. Details are now provided.

Theorem 10.5. *Given a piece A of a cake X such that $\mu_1(A) = 0$ and $\mu_2(A) = a$, $0 < a < 1$, there is a finite algorithm that produces exact fair division of X in the ratio $\alpha_1 : \alpha_2$, where $\alpha_i > 0$, $\alpha_1 + \alpha_2 = 1$.*

Proof: Set aside A and divide $X - A = Y_1 \cup Y_2$ in ϵ near-exact portions in the ratio $\alpha_1 - 2\epsilon : \alpha_2 + 2\epsilon$ (where ϵ will be chosen sufficiently small as described below). Then

$$\left| \frac{\mu_1(Y_1)}{\mu_1(X - A)} - (\alpha_1 - 2\epsilon) \right| = |\mu_1(Y_1) - (\alpha_1 - 2\epsilon)| < \epsilon,$$

so $\alpha_1 - 3\epsilon < \mu_1(Y_1) < \alpha_1 - \epsilon$.

It follows that $\mu_1(Y_2) \geq \alpha_2 + \epsilon$. Also $\left|\frac{\mu_2(Y_2)}{\mu_2(X - A)} - (\alpha_2 + 2\epsilon)\right| < \epsilon$ and, since $\mu_2(X - A) = 1 - a$, we have $(\alpha_2 + \epsilon)(1 - a) < \mu_2(Y_2) < (\alpha_2 + 3\epsilon)(1 - a)$. Thus, for ϵ sufficiently small we know $6\epsilon < \mu_2(Y_2) < \alpha_2$.

Now P_2 can divide Y_2 into 2 or more pieces all of which have μ_2 measure between 3ϵ and 6ϵ. At least one of the pieces, say Y_2', satisfies $\mu_1(Y_2') \geq 3\epsilon$ since $\mu_1(Y_2) > \alpha_2 > \mu_2(Y_2)$. Let P_1 cut a piece Y_2'' from Y_2' so that $\mu_1(Y_1 \cup Y_2'') = \alpha_1$. Furthermore, $\alpha_2 > \mu_2(Y_2 - Y_2'') > (\alpha_2 + \epsilon)(1 - a) - 6\epsilon = \alpha_2 - a(\epsilon + \alpha_2) - 5\epsilon > \alpha_2 - a$ for ϵ sufficiently small. This insures that P_2 can cut $A = A_1 \cup A_2$ so that $\mu_2\left((Y_2 - Y_2'') \cup A_1\right) = \alpha_2$ while $\mu_1(Y_1 \cup Y_2'' \cup A_2) = \alpha_1$.

We conclude this section by presenting another finite, unbounded algorithm given by Brams and Taylor for envy-free division of cake among n players [BT2].

The Brams and Taylor Envy-Free Algorithm

Although we present the Brams and Taylor Algorithm here for equal portions, it easily generalizes to unequal rational portions. We will describe the algorithm for four players, since that case illustrates the essential ideas and more players make that presentation more complicated.

In both envy-free algorithms, cuts are made by one player, and if by good fortune there is complete agreement among all players the task is quickly accomplished. Otherwise, two or more players disagree on an existing piece. In the algorithm of Theorem 10.3 that disagreement is managed with the tool of near-exact divisions, producing two smaller subcases where induction applies. In the Brams and Taylor Algorithm the disagreement is exploited by creating advantages players have over each other on certain portions of cake that the algorithm constructs. The cake is dispensed among the players in a finite sequence of envy-free assignments until the remaining portion is small enough not to destroy the advantages that have accrued along the way.

We start with some terminology and agreement on notation that will help simplify matters. We will write "Player P orders pieces $A < B < \alpha$" to mean "$\mu_P(A) < \mu_P(B) < \alpha$." Note that this does not require $A \subset B$. If players P_1 and P_2 have been assigned pieces A and B respectively, and R (a remainder portion) is a third piece, we will say P_1 *holds an R advantage over* P_2 if P_1 feels $A \geq B \cup R$. Note that holding an R advantage over another player depends on the assigned pieces A and B. If $\mu_1(A) - \mu_1(B) = \alpha > 0$, we will say P_1 holds an advantage of size α over P_2. If P_1 holds an R (or α) advantage over P_2 when A and B have been assigned as above, then if P_1 and P_2 receive further envy-free portions C and D respectively, then P_1 will still hold an R (or α) advantage over P_2 with respect to the assigned portions $A \cup C$ and $B \cup D$.

It is very useful if P_1 and P_2 can be assigned pieces A and B in such a way that each holds an advantage over the other of some size $\alpha > 0$, which is precisely what disagreement on a piece produces in the Brams and Taylor Algorithm. Recall that advantages were also at the heart of the Selfridge Envy-Free Algorithm (Section 1.4) and the Woodall Algorithm for strong fair division (Section 5.2).

We now present the two basic algorithms that are used repeatedly in the Brams and Taylor Algorithm.

Algorithm I: Given a piece of cake A, four players, and an $\epsilon > 0$, a subset B of A can be divided among the four players in an envy-free way so that the remaining portion $R = A - B$ has value smaller than ϵ to all four players.

Proceed as follows:

Step 1. Player P_1 cuts A into five equal pieces.

Step 2. Player P_2 identifies the three largest pieces and trims the larger two so that three equal pieces result. (We now have three pieces that P_1 cut and two pieces that P_2 may have trimmed; the trimmings are set aside.)

Step 3. Player P_3 identifies the two largest of the five pieces produced by Step 2. The larger is trimmed to produce two pieces of equal size.

Step 4. The players now choose four of the five pieces generated in Step 3 in the order P_4, P_3, P_2, P_1 with the restriction that both P_2 and P_3 must choose a piece they trimmed if it is available.

Note that P_4 chooses first, each of P_2 and P_3 will get one of their largest pieces, and P_1 will get an untrimmed piece. Hence those four pieces are distributed envy-free. Furthermore, P_1 views the remainder (which is unassigned) as having value at most 4/5 the value of A.

By repeating the four steps on the remainder repeatedly and by letting each player perform the role of P_1, the task is accomplished. (We also note that advantages are not utilized in Algorithm I.)

Algorithm II: If player P_1 views A and B as equal but P_2 views B as larger, then P_1 can be assigned part of A, P_2 can be assigned part of B, and P_3 and P_4 can be assigned parts from $A \cup B$ so that the division of the four pieces is envy-free. Furthermore, P_1 and P_2 will both hold an $\alpha > 0$ advantage over each other, where the size of α is a function of the size of the disagreement of P_2 on pieces A and B.

Let us assume that $\mu_2(B)/\mu_2(A) > ((10^k + 10)/10^k)$ for an appropriate positive integer k, and that in appropriate units $\mu_2(B) > 10^k + 10$ and $\mu_2(A) < 10^k$.

Step 1. Have P_1 cut A and B into 10^k+2 and 10^k+3 equal pieces respectively.

We now look for three pieces A_1, A_2, A_3 cut from A and three pieces B_1, B_2, B_3 from B so that P_2 orders them $A_1 \leq A_2 \leq A_3 < 1 < B_1 \leq B_2 \leq B_3$. (Appropriate trimming on these six pieces will enable four of them to be given out envy-free.) First note that at least three pieces cut from A must have value less than one to P_2; otherwise, there are at least 10^k of them with value at least one, contradicting $\mu_2(A) < 10^k$. Let P_2 order those pieces $A_1 \leq A_2 \leq A_3 < 1$.

If P_2 orders the pieces cut from B as $B_1 \geq B_2 \geq \cdots \geq B_{10^k+3}$ and if $B_3 > 1$, we have the six pieces $A_1 \leq A_2 \leq A_3 < 1 < B_3 \leq B_2 \leq B_1$ as desired. On the other hand, if $B_3 \leq 1$, then $\mu_2(B - (B_1 \cup B_2 \cup B_3)) \leq 10^k$ so that $3\mu_2(B_1) \geq \mu_2(B_1 \cup B_2 \cup B_3) \geq 10$ and $\mu_2(B_1) > 3$. Now P_2 can cut B_1 into three pieces $B_1' \cup B_2' \cup B_3'$ with $A_1 \leq A_2 \leq A_3 < 1 < B_1' \leq B_2' \leq B_3'$ as desired.

So we may now assume six pieces have been found, three from A and three from B, that P_2 orders $A_1 \leq A_2 \leq A_3 < 1 < B_1 \leq B_2 \leq B_3$, while P_1 views the three pieces from A as equal and all strictly larger than the three pieces from B since P_1 views A and B as equal and A is cut into fewer pieces.

Step 2. Player P_3 identifies the two largest pieces of the six and trims the larger so that the two are equal in size.

Step 3. Four of the six pieces resulting from Step 2 are chosen in the order P_4, P_3, P_2, P_1 with the condition that P_3 must take the piece P_3 trimmed if P_4 doesn't take it.

It is readily checked that there is no envy on the four pieces and that P_2 and P_1 will have chosen one of the B_i's and A_i's respectively. So for an appropriate $\alpha > 0$ each holds an α advantage over the other. Thus, the claim made in the statement of Algorithm II is justified.

We can now proceed with the envy-free division of the entire cake X among P_1, P_2, P_3, and P_4. First P_1 cuts X into four equal pieces. If there is total agreement, we are done. Otherwise, there is a pair, say P_1 and P_2, so that Algorithm II can be used to dispense of part of the cake in an envy-free manner while creating some reciprocal α_1 advantage among P_1 and P_2. Continue to dispense cake envy-free using Algorithm I until the remainder R has value smaller than α_1 to all the players.

Next have P_1 cut R into four equal parts. If both P_3 and P_4 agree with those cuts, R can be given out envy-free (even when P_2 disagrees, P_2 chooses first) and we are done. Otherwise, Algorithm II can be reapplied to the pair P_1 and (say) P_3, to create a mutual α_2 advantage between them.

If a stage is reached where three players all hold a reciprocal advantage over a fourth, that fourth player can be given all of the remainder (after it is reduced properly using Algorithm I), so all of the cake is given out envy-free. Otherwise, the general step is to let any player cut the remainder into 12 equal portions. Divide the four players into two sets, those who agree with the cutter and those who don't.

Case 1: If there are two players, one from each set of players, who do not yet hold reciprocal advantages over each other, apply Algorithms II and I (in that order) to create a reciprocal advantage between the pair and then reduce the size of the remainder to a portion smaller than all established advantages.

Case 2: If Case 1 does not hold, let each player who agrees with the cutter (including the cutter) take an equal number of the 12 available pieces. In this case all of the cake is assigned to the four players in an envy-free manner, since all players who did not get a final portion hold an advantage over all players who did get a final portion that is larger than all of these remaining portions.

It is clear, if Case 2 is never encountered, that eventually some three players will all hold an advantage over the fourth and the algorithm terminates.

10.4 Super Envy-Free Division

An even stronger type of envy-free division among n players has been studied by Barbanel.

> *The cake division $X = X_1 \cup \cdots \cup X_n$*
> *is super envy-free whenever $\mu_i(X_j) < 1/n$ for $i \neq j$.*

In this division, each player thinks every other player has received less than a fair share, and therefore each feels he or she has more than a fair share, i.e., $\mu_i(X_i) > 1/n$. Of course this type of division is not always possible, for example, when all n measures are the same. As it turns out, the existence of such a division depends completely on whether or not the measures are linearly independent.

The notion of linear dependence of measures is a natural one. Two measures μ_1 and μ_2 are dependent only when $\mu_1 = \mu_2$, that is, they assign the same value to each measurable set. If μ_1 and μ_2 are independent (i.e., not equal) and α_1 and α_2 are two positive numbers with $\alpha_1 + \alpha_2 = 1$, a third measure μ_3 can be defined by $\mu_3(X) = \alpha_1\mu_1(X) + \alpha_2\mu_2(X)$ for each measurable set X (assumed to be the same class of sets for all measures). In this case, the measures μ_1, μ_2, and μ_3 are dependent.

So the definition of independence of a set of measures is a familiar one:

Probability measures $\mu_1, \mu_2, \cdots, \mu_n$ are linearly independent if and only if for each i, and for any choice of scalars $\alpha_1, \cdots, \alpha_{i-1}, \alpha_{i+1} \cdots, \alpha_n$, it is not the case that $\mu_i = \alpha_1\mu_1 + \cdots + \alpha_{i-1}\mu_{i-1} + \alpha_{i+1}\mu_{i+1} + \cdots + \alpha_n\mu_n$.

The theorem of Barbanel is the following [Bar1]:

Theorem 10.6. *A super envy-free partition $X = X_1 \cup \cdots \cup X_n$ exists if and only if the measures $\mu_1, \mu_2, \cdots, \mu_n$ are linearly independent.*

It is natural to ask whether an algorithm can produce such a super envy-free division. In some situations the answer is yes, as the following algorithm due to Webb shows [Web4].

An Algorithm for Super Envy-Free Division

In order to accomplish super envy-free division the players must use linearly independent measures. How can we know if their measures are independent, and even if that is the case how are super envy-free portions found? There is a situation that could arise where both questions can be resolved. Suppose there are n disjoint subsets $X_1, \cdots, X_n$ of the cake X so that the value matrix $V = (v_{ij}) = (\mu_i(X_j))$ is non-singular. This would certainly mean the measures are independent. (The term *witness* has been used for pieces of cake and their values assigned by different measures certifying that these measures are not equal. Hence, a set A with $\mu_1(A) \neq \mu_2(A)$ is a witness that $\mu_1 \neq \mu_2$. The pieces $X_1, \cdots, X_n$ and matrix V are a witness that the measures are linearly independent. We have seen several algorithms using some witness in their description.)

We first prove that we may assume $X = X_1 \cup \cdots \cup X_n$; i.e., the entire cake is used to produce the witness V. Since V is non-singular there is a solution to the equation $VY = [1, 1, \cdots, 1]^T$. Then the solution vector Y has a non-zero entry, say y_1, so that from Cramer's Rule we know

$$0 \neq \det \begin{bmatrix} 1 & v_{12} \cdots v_{1n} \\ \vdots & \\ 1 & v_{n2} \cdots v_{nn} \end{bmatrix} = \det \begin{bmatrix} 1 - v_{12} - \cdots - v_{1n} & v_{12} \cdots v_{1n} \\ \\ 1 - v_{n2} - \cdots - v_{nn} & v_{n2} \cdots v_{nn} \end{bmatrix}.$$

So from the pieces generating the original matrix V we simply replace X_1 by the complement of $X_2 \cup \cdots \cup X_n$. Let us thus assume that $X = X_1 \cup X_2 \cup \cdots \cup X_n$ in what follows.

The matrix V is stochastic (all row sums are one, with non-negative entries) and V^{-1} also has row sums of one (although it may have negative entries). Set matrix $N = [n_{ij}]$ where

$$n_{ij} = \begin{cases} (1/n) + \delta & i = j \\ (1/n) - \delta/(n-1) & i \neq j \end{cases}$$

for an appropriately small $\delta > 0$ chosen so that $V^{-1}N = R = (r_{ij})$ is stochastic. (Row sums of V^{-1} are one and columns of N can be made close enough to $(1/n)(1, 1, \cdots, 1)^T$ so that all r_{ij} are non-negative.)

Applying the near-exact division of Theorem 10.2, we have a finite, bounded (in terms of ϵ) algorithm that gives an ϵ near-exact division of piece $X_j = Y_{j1} \cup \cdots \cup Y_{jn}$ in the ratios $r_{j1} : \cdots : r_{jn}$ for each $j = 1, 2, \cdots, n$. Then $\mu_i(Y_{jk}) = v_{ij}r_{jk} + \lambda_{ijk}$ where $|\lambda_{ijk}| < \epsilon$.

Dividing all of the pieces $X_1, X_2, \cdots, X_n$ accordingly and giving P_k portion $Y_{1k} \cup \cdots \cup Y_{nk}$ we can check that for $i \neq k$ we get $\mu_i(Y_{1k} \cup \cdots \cup Y_{nk}) = (v_{i1}r_{1k} + \lambda_{i1k}) + \cdots + (v_{in}r_{nk} + \lambda_{ink}) = n_{ik} + (\lambda_{i1k} + \cdots + \lambda_{ink}) < 1/n$ provided $\epsilon < \delta/n^2$.

Thus, super envy-free division has been accomplished.

10.5 EXERCISES

10.1. Exact division for two players in the ratio $\alpha_1 : \alpha_2$ can be accomplished more easily than it was in Theorem 10.5 if more is assumed. Specifically, assume there is a piece A with $\mu_1(A) = a > 0, \mu_2(A) = 0$ and a piece B with $\mu_1(B) = 0, \mu_2(B) = b > 0$.

(a) Show we may assume A and B are disjoint without loss of generality.

(b) Give an algorithm using near-exact division and the pieces A and B that accomplish exact division in the ratio $\alpha_1 : \alpha_2$.

10.2. Extend the ideas in Exercise 1 to n players.

10.6 PROJECTS

10.1. Can Theorem 10.5 be extended to more players with weaker assumptions than those used in Exercises 1 and 2?

10.2. The main unresolved issue in this section is the general question of finding bounded finite algorithms for envy-free division for any number of players. This is a well-known problem, and any algorithm for envy-free division, even for special small cases and even using continuous moving knife methods, beyond those presented in Chapters 5 and 10 would be of interest.

ELEVEN

A Return to Division for Unequal Shares

11.1 Resetting the Stage

In Sections 3.2 and 3.3 some algorithms for fair division were presented in which the shares are not equal. At that point, we made some statements whose justifications were postponed. It is now time to return to some of those details.

There are three specific tasks. First we will develop the theory of fair division based on Ramsey partitions as introduced in Section 3.2. Next we will justify that the infinite discrete division, based on binary representations of the size of the shares (which are allowed to be irrational), does in fact accomplish the fair division required. Finally, we will see how unequal shares affect envy-free division.

11.2 Ramsey Partitions and Fair Division

If two players are to share a cake in the ratio $a : b$ where a and b are relatively prime positive integers, we saw in Section 3.2 that there are some special Ramsey partitions of the integer $a + b$ which are nicely suited to the task. Recall the definition [MRW].

If a and b are relatively prime positive integers, the partition $\mathcal{P} = (p_1, p_2, \cdots, p_n)$ of $a+b$ is called a *Ramsey partition* for the pair a, b provided:

(1) $\sum_{i=1}^{n} p_i = a + b$, and

(2) If S is any subset of $\mathcal{P}$, either:

(i) $\sum_{S'} p_i = a$ for some $S' \subset S$, or

(ii) $\sum_{S''} p_i = b$ for some $S'' \subset \mathcal{P} - S$.

For example, $\mathcal{P} = (8, 5, 3, 2, 1, 1, 1)$ is Ramsey for the pair 8, 13. Table 3.1 in Section 3.2 lists the other 45 Ramsey partitions for the pair 8, 13. Recall further that the algorithm then calls for one player to cut the cake in pieces whose sizes are in the ratios $p_1 : p_2 \cdots : p_n$ and the non-cutter to circle the pieces he or she would judge to be at least as large as the cutter thinks them to be. If we call these circled pieces "acceptable" (to the non-cutter), then, because the partition is Ramsey, either the non-cutter can get his or her fair share from acceptable pieces or the cutter can get his or her fair share from unacceptable pieces. In the latter case, the non-cutter is giving away a portion of the cake that is considered smaller than the cutter's share. In both cases, the cutter receives pieces that total exactly the cutter's share. In either case, fair division in the ratio $a : b$ is easily accomplished.

We turn now to developing some of the number theoretic aspects of Ramsey partitions.

In all that follows, we will assume unless otherwise noted that the numbers in any partition are listed in descending order of size.

We first state some tests for Ramsey partitions.

Theorem 11.1. *The following are equivalent for the partition* $\mathcal{P} = (p_1, p_2, \cdots, p_n)$.

(1) $\mathcal{P}$ *is Ramsey for the pair* a, b.

(2) For any $S \subset \mathcal{P}$, $\sum_{S} p_i \geq a$ *implies for some* $S' \subset S$ *that* $\sum_{S'} p_i = a$,

and for any $S \subset \mathcal{P}$, $\sum_{S} p_i \geq b$ *implies for some* $S' \subset S$ *that* $\sum_{S'} p_i = b$.

(3) *For any* $i_1 < i_2 < \cdots < i_t$ *satisfying* $p_{i_1} + p_{i_2} + \cdots + p_{i_{t-1}} < a$ *and* $p_{i_1} + p_{i_2} + \cdots + p_{i_t} \geq a$, *we must have* $p_{i_1} + p_{i_2} + \cdots + p_{i_t} = a$. *A similar statement holds for* b.

Proof: We show (1) $\Rightarrow$ (2) $\Rightarrow$ (3) $\Rightarrow$ (1). If $\sum_{S} p_i > a$ and $\sum_{S'} p_i \neq a$ for all $S' \subset S$, then neither of the conditions in (2) of the definition for Ramsey partitions is met since $\sum_{\mathcal{P}-S} p_i < b$. So (1) $\Rightarrow$ (2).

If $p_{i_1} + p_{i_2} + \cdots + p_{i_{t-1}} < a$ while $p_{i_1} + p_{i_2} + \cdots + p_{i_t} > a$, then (2) fails since the terms are listed in descending order. So (2) $\Rightarrow$ (3). Finally, the Ramsey condition follows immediately from (3) since, for any S, either $\sum_{S} p_i \geq a$ or $\sum_{\mathcal{P}-S} p_i \geq b$.

Thus, properties that Ramsey partitions must have include:

1. No term in $\mathcal{P}$ exceeds a or b.

2. $\mathcal{P}$ does not contain exactly $a-1$ or $b-1$ ones.

3. For some j_1, $(p_1, p_2, \cdots, p_{j_1})$ is a partition of a and $(p_{j_1+1}, \cdots, p_n)$ is a partition of b. Similarly, for some j_2, $(p_1, p_2, \cdots, p_{j_2})$ partitions b and $(p_{j_2+1}, \cdots, p_n)$ partitions a.

4. Assuming $a < b$, with j_1 as in 3, $P' = (a, p_{j_1+1} \cdots, p_n)$ is also Ramsey for a, b.

The next result will be used to construct the unique Ramsey partition for a, b with the fewest terms, which we will call the *minimal Ramsey partition* for a, b.

Theorem 11.2. *Let* $p_1 < a < b$ *and* $\mathcal{P} = (p_1, p_2, \cdots, p_n)$ *partition* b. *Then* $\mathcal{P}' = (a, p_1, p_2, \cdots, p_n)$ *partitions* $a + b$ *and:*

(1) $\mathcal{P}'$ *is Ramsey for* a, b *if and only if* $\mathcal{P}$ *is Ramsey for* $a, b - a$, *and*

(2) $\mathcal{P}'$ *is minimal Ramsey for* a, b *if and only if* $\mathcal{P}$ *is minimal Ramsey for* $a, b - a$.

Before giving a proof, let us use the result to construct a minimal Ramsey partition.

Example: Let us construct the minimal Ramsey partition for the pair 7,12. From 1 and 4 above, the partition we are after must start with a 7, and from

Theorem 11.2, the 7 is followed by the minimal Ramsey partition for 7,5. Applying the argument again, we then need a 5 followed by the minimal Ramsey partition for 5,2. Next comes a 2 followed by the minimal Ramsey partition for (2,3). Continuing in this way, the desired minimal Ramsey partition for 7,12 is $\mathcal{P} = (7, 5, 2, 2, 1, 1, 1)$.

Proof of Theorem 11.2:

For (1), if $\mathcal{P}'$ is Ramsey for a, b and if in $\mathcal{P}$ we have $p_{i_1} + p_{i_2} + \cdots + p_{i_{t-1}} < b - a$ and $p_{i_1} + p_{i_2} + \cdots + p_{i_t} \geq b - a$, then $a + p_{i_1} + p_{i_2} + \cdots + p_{i_{t-1}} < b$ and $a + p_{i_1} + p_{i_2} + \cdots + p_{i_t} \geq b$, which requires $p_{i_1} + p_{i_2} + \cdots + p_{i_t} = b - a$ by Theorem 11.1 (3). So sums in $\mathcal{P}$ cannot skip $b - a$, and it is clear they cannot skip a because $\mathcal{P}'$ is Ramsey for a, b. Thus, $\mathcal{P}$ is Ramsey for $a, b - a$.

Conversely, if $\mathcal{P}$ is Ramsey for $a, b - a$ and if sums from $\mathcal{P}'$ skip b, since $p_1 + p_2 + \cdots + p_n = b$, we must have $a + p_{i_1} + p_{i_2} + \cdots + p_{i_t} < b$ and $a + p_{i_1} + p_{i_2} + \cdots + p_{i_t} > b$. Subtracting a from both inequalities leads to a contradiction that $\mathcal{P}$ is Ramsey for $a, b - a$.

For (2), in order for $\mathcal{P}$ to be minimal Ramsey for a, b, the first term of $\mathcal{P}$ must be a. It follows from (1) that $\mathcal{P}$ and $\mathcal{P}'$ are both minimal or both non-minimal.

Let us now formalize the construction. In our fair division applications, we will always assume $gcd(a, b) = 1$.

> **Construction of the Minimal Ramsey Partition for the Pair *a,b*, Where $a < b$:**
>
> **Step 1.** Choose the first part equal to a.
>
> **Step 2.** If $a = b - a = 1$, the last three terms are $1, 1, 1$.
>
> **Step 3.** If $b - a \neq 1$, repeat with the new ratio $a : b - a$.

As we have noted with an example in Section 3.2, the method of construction for the minimal Ramsey partition for a, b gives the following:

> The number of terms in the minimal Ramsey partition for a, b is one more than the sum of the quotients in the Euclidean Algorithm for determining $gcd(a, b)$.

The main point in using the minimal Ramsey partition, rather than all ones, when doing fair division in the ratio $a : b$ is to be more efficient and use fewer cuts. So it is natural to ask how efficient we are being. Obviously cutting all ones produces $a + b$ pieces. Also, if $a = 1$ then all ones is the minimal (and only) Ramsey partition. In all other cases we get significant improvement. Even for $a = 2$ and b odd, the minimal Ramsey partition is $(2, 2, \cdots, 2, 1, 1, 1)$ having $(b + 1)/5$ parts. We will return to the general case below.

In addition to examining the general efficiency of divisions using Ramsey partitions, we can raise the question of its comparison with the efficiency of the Cut Near-Halves Algorithm. In Section 3.3 the claim was made that division in the ratio $a : b$, $gcd(a, b) = 1$, using Cut Near-Halves would require at most r cuts, where $2^{r-1} < a + b \leq 2^r$. Note that when a and b are both odd and $a + b = 2^r$, the sizes of the pieces both cut and chosen at each step are $2^{r-1}, 2^{r-2}, \cdots, 2, 1$ so that r cuts are required. Similarly, if $a + b = 2^{r-1} + 1$, then if the smaller piece is chosen at each step, the sizes of successive pieces to be further divided are $2^{r-2}+1, 2^{r-3}+1, \cdots, 3, 2, 1$ and again the number of cuts required is r. Finally, if $2^{r-1} + 1 \leq a + b \leq 2^r$, then the sizes of the pieces to be further divided at each step are at most $2^{r-1}, 2^{r-2}, \cdots, 2, 1$ so r cuts always suffice. Examples where not all r cuts are required are found in Section 3.3. Thus for any ratio $a : b$, $gcd(a, b) = 1$, the Cut Near-Halves Algorithm uses no more than $\lceil \log_2(a + b) \rceil$ cuts.

Could the Ramsey partition method ever be more efficient than the Cut Near-Halves Algorithm? Suppose that for some $a > b \geq 1$, $gcd(a, b) = 1$, the Ramsey partition uses $s - 1$ cuts (generating s pieces), but the Cut Near-Halves Algorithm requires at least s cuts. Then we have just seen that $a + b > 2^{s-1}$ so $a > 2^{s-2}$. We can show that if the Ramsey partition has s parts, then $a \leq F_s$, where F_n denotes the nth Fibonacci number [LoWe]. (Note that the minimal Ramsey partition for the ratio $F_s : F_{s-1}$ is $(F_{s-1}, F_{s-2}, \cdots, F_1, 1)$, having s parts.) Now it is easy to check that $F_s \leq 2^{s-2}$. This is true for $s = 2$ and 3, and if $F_{s-2} \leq 2^{s-4}$ and $F_{s-1} \leq 2^{s-3}$ then $F_s = F_{s-2} + F_{s-1} \leq 2^{s-4} + 2^{s-3} \leq 2^{s-2}$. Thus the Ramsey Partition Algorithm is never more efficient than the Cut Near-Halves Algorithm. Roughly speaking, the number of cuts for Cut Near-Halves grows like $\log_2 a$ or smaller, while the number of cuts for Ramsey partitions grows like $\log_\varphi a$ or larger, where $\varphi = (1 + \sqrt{5})/2$.

Finally, we note that there is an asymptotic count for the number of Ramsey partitions $N(k, nk + r)$ for the pair $k, nk+r$, where $0 \leq r < k$ [RW5]. Specifically, $N(k, nk + r)$ is asymptotic to $cn^{\Omega(k)}$, where c is a constant independent of the residue class r and $\Omega(k)$ is the number of prime factors (counting repetitions) of k. Only in very special cases has c been explicitly given. Details are quite involved and too far removed from our focus on fair division algorithms to justify their inclusion.

11.3 Cut and Choose for Unequal Shares

In Section 3.3 the Cut Near-Halves Algorithm was described. Because the ratio of shares was rational, the algorithm was finite bounded, and the justification that it accomplished the desired division was straightforward. However, we see that both the near-halves and Ramsey partition procedures make no sense for irrational ratios. So algorithms based on those notions do not apply for irrational portions. But we do already know a finite, unbounded algorithm that will accomplish fair division for unequal ratios, rational or irrational. In fact, the envy-free, near-exact algorithm presented in Section 10.3 accomplished a division that was not only fair, but also envy-free and near-exact for any ratios and any number of players.

In Section 3.5 an algorithm that is an extension of simple Cut and Choose and based on the binary representation of the shares was briefly introduced. In some rational and all irrational cases, those binary representations are non-terminating, so the algorithm, while discrete, is infinite. Nevertheless, the algorithm has one nice simple characteristic — every step is just the basic Cut and Choose. We will describe that algorithm again, justify that it does in fact accomplish the desired division, and examine some of the issues raised in the process.

Suppose the cake is to be divided in the ratio $a : b$, rational or irrational, with $a + b = 1$. Then both a and b have terminating binary representations or neither does. (Of course terminating forms also have non-terminating forms, e.g. $.1 = .0111\cdots$). Suppose $a = .a_1a_2a_3\cdots$ and $b = .b_1b_2b_3\cdots$ in binary representation. If these terminate in the nth place, then $a_i + b_i = 1$ whenever $1 \leq i < n$ and $a_n = b_n = 1$. In the case neither terminates, $a_i + b_i = 1$ for all $i \geq 1$. We are now prepared to formally state the algorithm.

Repeated Cut and Choose Algorithm for Dividing a Cake in the Ratio $a : b$

Step 1. With $a + b = 1$ write $a = .a_1a_2a_3$ and $b = .b_1b_2b_3\cdots$ in binary form, using terminating forms if possible.

Step 2. At Step i, P_1 cuts halves if $a_i = 0$; P_2 cuts halves if $b_i = 0$; either cuts if $a_i = b_i = 1$. The other chooses and claims one of the two pieces.

Step 3. The procedure stops when $a_i = b_i = 1$ with the unclaimed piece given to the cutter. Otherwise, the procedure continues on the unclaimed piece.

An example of a terminating case is found in Section 3.5, and for this finite case the justification that the pieces are fair is very similar to that found in Section 3.3. Recall that the argument went something like the following. After the first step, the over-all division is fair if the later division starting from Step 2 is fair on the unclaimed piece for appropriate adjusted ratio. But that division in turn will be fair if the division from Step 3 onward is fair on the unclaimed piece (after Step 2) in yet another adjusted ratio. In the finite case, the buck is passed n times until the question of fairness for the entire process reduces to whether or not Cut and Choose will guarantee each player at least half of the last unclaimed piece. Therefore, we know over-all fair division in the original ratios on the entire cake has been accomplished.

But now we face a new problem for the non-terminating case — there is no final step in which to corner the fair issue. So new methods of justification are required, and that is our next order of business.

Let us suppose Tom and Dick are to receive portions of size a and b respectively. Through k steps, they each will have chosen and claimed certain portions of the cake. (One player may have been awarded all of the first k pieces if that player has digits all ones through the first k places. Recall that the cutter at Step i is the player whose ith digit is 0, but the chooser has the ith digit 1.) We will denote these two total claimed holdings after k steps by T_k and D_k respectively, and the remaining unclaimed portion will be denoted by R_k. Also, set $A_k = .a_1a_2\cdots a_k$ and $B_k = .b_1b_2\cdots b_k$.

We show that through k steps both Tom and Dick think they are doing fine so far and that the other is not ahead of the game. Specifically, for $k = 1, 2, 3\cdots$ we will have:

(i) $\mu_T(T_k) \geq A_k$,

(ii) $\mu_T(D_k) \leq B_k$,

(iii) $\mu_D(T_k) \leq A_k$,

(iv) $\mu_D(D_k) \geq B_k$,

(v) $\mu_T(R_k) \leq 1/2^k$ and $\mu_D(R_k) \leq 1/2^k$.

The fact that fair division is accomplished in the non-terminating case follows immediately by letting $k \to \infty$ in (i) and (v). The case for terminating forms after n digits is even more apparent, because in this case $a = A_n$ and $b = B_n$.

The justification for the five statements is by induction on k. For $k = 1$, the chooser will think he has received at least half of the cake and the cutter will think exactly half has been claimed by the chooser. Statements (i) - (v) are now clear for $k = 1$. Assuming the kth case, what happens at step $(k + 1)$?

Case 1: $a_k = 0, b_k = 1$.

Tom will cut R_k in two pieces and Dick will choose and claim one of the two pieces. We know $\mu(R_k) = 1 - \mu(D_k) - \mu(T_k)$ for both measures. Since $T_k = T_{k+1}, \mu_T(T_{k+1}) = \mu_T(T_k) \geq A_k = A_{k+1}$ from the induction assumption (i), and $\mu_D(T_{k+1}) = \mu_D(T_k) \leq A_k = A_{k+1}$ from (iii).

Also $\mu_D(D_{k+1}) \geq \mu_D(D_k) + (1/2)(1 - \mu_D(T_k) - \mu_D(D_k)) = (1/2)(1 + \mu_D(D_k) - \mu_D(T_k)) \geq (1/2)(1 + B_k - A_k) = (1/2) + B_k - (1/2)(B_k + A_k) = (1/2) + B_k - (1/2)(1 - 1/2^k) = B_k + 1/2^{k+1} = B_{k+1}$. For (ii), $\mu_T(D_{k+1}) = \mu_T(D_k) + (1/2)(1 - \mu_T(T_k) - \mu_T(D_k)) = (1/2)(1 + \mu_T(D_k) - \mu_T(T_k)) \leq (1/2)(1 + B_k - A_k) = B_{k+1}$.

Finally, for $(v), \mu(R_{k+1}) \leq (1/2)\mu(R_k) \leq (1/2)(1/2^k) = 1/2^{k+1}$ for both measures.

Case 2: $a_{k+1} = 1, b_{k+1} = 0$.
This is established the same way as Case 1.

Case 3: $a_{k+1} = b_{k+1} = 1$.
This case is established by observing that R_{k+1} is empty, $A_{k+1} + B_{k+1} = 1$, and by justifying (ii) and (iv) exactly as in Case 1.

The proof is now complete and fair division has been accomplished.

Moreover, this algorithm is the only sequence of cut-and-choose steps that can guarantee fair shares in the ratio of $a : b$. As long as $a_i + b_i = 1$, the cutter must be the player corresponding to the 0 entry. It is certainly possible that there is agreement through the previous $i - 1$ steps, so that $\mu_T(T_{i-1}) = \mu_D(T_{i-1}) = A_{i-1}$ and $\mu_T(D_{i-1}) = \mu_D(D_{i-1}) = B_{i-1}$. If $a_i = 1$ and Tom cuts, Dick can claim at least half of what is left (according to Tom's measure μ_T), so that $\mu_T(D_i) \geq B_{i-1} + 1/2^i$. But then $a + \mu_T(D_i) > 1$, so Tom cannot receive a fair piece. It follows that there is a finite cut-and-choose procedure if and only if the binary forms for a and b terminate.

We now consider more than two players using the methods above. Some interesting things happen.

Example 2: Suppose Tom, Dick and Harry are to share in the ratios $1/2 : 1/3 : 1/6$. Using Repeated Cut and Choose, Tom and Dick can first divide the cake in the ratio $1/2 : 1/3$, which is $3/5 : 2/5$. Then Harry can repeat the algorithm in the ratio $1/6 : 5/6$ with each of Tom and Dick. Then Tom will have $(5/6)(3/5) = 1/2$, Dick will have $(5/6)(2/5) = 1/3$, and Harry will have 1/6th of everything. In order to avoid a sequence of three infinite procedures back-to-back, a single countable process can be described using a diagonalization procedure that permits Harry to start the division of pieces with Tom and Dick as soon as they are claimed by Tom or Dick. Before the next Tom-Dick step is performed, perform a Tom-Harry and Dick-Harry step on every piece Tom or Dick may have claimed in the Tom-Dick division. Once a piece is claimed in either the Tom-Harry or Dick-Harry division, it remains uncut.

For $P_1, \cdots, P_n$ to share in the ratios $a_1 : a_2 : \cdots : a_n$, first have $P_1, \cdots, P_{n-1}$ share in the ratios $a_1/(1 - a_n) : a_2/(1 - a_n) : \cdots : a_{n-1}/(1 - a_n)$. Then P_n

will share with each of $P_1, \cdots, P_{n-1}$ in the ratio $a_n : 1 - a_n$. The procedures can be diagonalized so that a single countable sequence suffices.

When is a finite procedure possible? For $n = 2$ this is an easy question; the procedure is finite if and only if $a/(a+b) = p/2^m$ for some positive integers p and m. Hence, the ratio $5 : 11$ is accomplished by 4 steps, whereas the ratio $1/\pi : (\pi - 1)/\pi$ requires an infinite procedure.

The case for three or more players is more interesting. For example, suppose the ratios are $12 : 3 : 1$. If Tom and Dick divide first, the ratio is $4/5 : 1/5$, leading to an infinite process. Yet, if Dick and Harry divide first in the ratio $3/4 : 1/4$, that process is finite and must be followed by Tom dividing with each of Dick and Harry in the ratio $3/4 : 1/4$, each of which is finite. In general, the inductive procedure given above for n players is finite if and only if for some permutation $a_1 : a_2 : \cdots : a_n$ of the ratios, each of the fractions $a_i/(a_1 + a_2 + \cdots + a_i)$, $2 \leq i \leq n$, can be written as $p_i/2^m$.

The example above shows that the order of the divisions can be important. Indeed, for $n \geq 3$, the procedure cannot be finite for *all* possible permutations of the ratios. Suppose we have three positive numbers a, b, c with $a+b+c=1$ so that the over-all process is finite regardless which pair goes first. This would require $a/(a+b)$ and $b/(a+b)$ to have the form $p/2^m$ and $q/2^m$, where $p+q=2^m$. Thus, $a/b = p/q$ is rational and similarly so is a/c. It follows that we can assume $a : b : c$ are the same ratios as $n_1 : n_2 : n_3$, where all n_i are integers and $gcd(n_1, n_2, n_3) = 1$.

By considering different orderings and stages of the divisions, we see that each of the fractions $n_1/(n_1+n_2)$, $n_1/(n_1+n_3)$, and $(n_1+n_2)/(n_1+n_2+n_3)$ must be reducible to a fraction of the form $p/2^n$ where p is odd and $n \geq 1$. In particular, this means n_1+n_2, n_1+n_3, and $n_1+n_2+n_3$ must all be even. But this can happen only if all of the numbers n_1, n_2, and n_3 are even, contradicting the fact that $gcd(n_1, n_2, n_3) = 1$.

11.4 Envy-Free Division for Three Players with Unequal Shares

Let's revisit envy-free division and see what can be done with a finite bounded algorithm if the players are to receive unequal shares. Since we still don't know of any such algorithm for four or more players even with equal shares, it seems our best hope is for only two or three players. If we use the idea of "cloning" as in Section 3.1, for unequal shares, other than $2 : 1$, cloning would require at least four players and we would seem to need the unbounded algorithms from Chapter 10.

Are there other ways to approach this problem? Fortunately, with some effort the algorithm given in Section 1.4, which awards three players (equal) envy-free portions, can be extended to cover the case where the players are to receive *unequal* rational portions [Web3]. Let us assume Tom, Dick, and Harry are to receive portions in the ratios $t : d : h$ where t, d, and h are positive integers with $gcd(t, d, h) = 1$. Recalling definitions found in Section 10.3, if $X = T \cup D \cup H$ then Tom does not envy Dick or Harry provided $\frac{\mu_t(T)}{t} \geq \frac{\mu_t(D)}{d}$ and $\frac{\mu_t(T)}{t} \geq \frac{\mu_T(H)}{h}$. The goal is to allocate portions so that none of the three envies another.

The procedure will rely on basic notions used in the equal shares Selfridge Envy-Free Algorithm. A first cutter, say Tom, will cut $t+d+h$ equal pieces. If the other two agree on those cuts, the task is simple. If they don't, the original pieces are ordered in size by Dick and some of the larger ones are trimmed. The trimmings are temporarily set aside. The first goal is to assign $t+d+h$ trimmed pieces, Tom getting t of them, Dick getting d, and Harry h, so that the players are envy-free on that assignment. Another critical feature of that division is that if Dick (or Harry) receives a trimmed piece, Tom would remain envy-free if all the cake trimmed from that piece were later awarded to Dick (or Harry). This advantage that Tom has is exploited in the division of the trimmings. Unfortunately, in the unequal shares case, the algorithm may require Harry also to get into the trimming act. In some cases he will have to restore some of the cake Dick has trimmed from pieces in order to make the envy-free assignments of the $t+d+h$ pieces.

Before looking at the general case, we look at a simpler case where the ratios are $2 : 2 : 1$ for Tom, Dick, and Harry respectively, and trimming by Harry is not required.

Step 1: Let Tom cut X into five equal pieces $X_1, \cdots, X_5$.

Step 2: Dick arranges the pieces in decreasing order of size, which we may assume is $X_1, \cdots, X_5$. He now trims off excess pieces E_1 and E_2 from X_1 and X_2 so that $X_1^{'} = X_1 - E_1$, $X_2^{'} = X_2 - E_2$, and X_3 are equal. (Possibly E_2 and even E_1 could be empty if Dick thinks X_1, X_2, and X_3 were equal to begin with.) Also the three pieces $X_1^{'}$, $X_2^{'}$, and X_3 are all at least as big as X_4 and X_5 to Dick. We will say that these three pieces are *acceptable* to Dick.

Step 3: Harry chooses any one of the pieces $X_1^{'}$, $X_2^{'}$, X_3, X_4, X_5. If Harry chooses one of the first three, Dick gets the other two. Otherwise, Dick gets $X_1^{'}$ and $X_2^{'}$. (In either case the two trimmed pieces $X_1^{'}$ and $X_2^{'}$ have been taken.) Tom gets the two remaining pieces. There are two important points to note so far:

(i) The distribution of pieces in Step 3 is envy-free in the ratio $2 : 2 : 1$, since each person believes that each piece he has received so far is at least as large as any other piece.

(ii) Using terminology introduced in Section 10.3, we see that Tom has an E_1 advantage with respect to $X_1^{'}$ over the player receiving $X_1^{'}$ (meaning if all of E_1 were given to that player, Tom would remain envy-free). Similarly he has an E_2 advantage with respect to $X_2^{'}$ over the player receiving that piece.

We use Tom's advantages to divide E_1 and E_2. The process is the same on both. Assume Dick has received $X_1^{'}$.

Step 4: Let Harry cut E_1 into five equal pieces. Dick chooses any two pieces. Tom chooses two of the remaining three pieces. Harry gets the fifth piece. Piece E_2 is similarly divided with obvious modifications depending on who received $X_2^{'}$.

It is easy to see that this distribution, together with the distribution in Step 3, is envy-free on X in the ratio $2:2:1$. Dick got the first two choices on E_1. So he receives two of the largest pieces in both Steps 3 and 4. Harry thinks all five pieces of E_1 are equal, so he too receives one of the largest pieces in both Steps 3 and 4. Tom chose before Harry on E_1 and had an E_1 advantage over Dick on piece $X_1^{'}$, so he too is envy-free after E_1 is divided. The argument on E_2 is identical.

Note that at most 14 cuts are needed.

To see how the general case can be more complicated, consider the ratios $5:4:3$.

Suppose Tom cuts 12 equal pieces and Dick trims the 3 biggest to equal the 4th biggest. Now if Harry chooses his 3 favorites, they could include some of the ones Dick trimmed. There aren't enough pieces acceptable to Harry and Dick. (There must be at least 7.) Maybe we could solve this problem by having Dick trim more than 3 pieces to begin with. If Dick trims the 6 biggest to equal the 7th biggest, then no matter which pieces Harry wants, there will be at least 7 acceptable to one or the other. But what if Harry chooses 3 of the untrimmed pieces? There are plenty of pieces left that are acceptable to Dick, but there are only 3 untrimmed pieces left and Tom has to get 5 pieces. He won't be envy-free of Harry if he has to take some trimmed pieces. These complications are addressed in what follows.

We will assume $t \geq d \geq h$ and $\gcd(t, d, h) = 1$. We begin by letting Tom cut $t + d + h$ equal pieces. Then Dick trims some pieces and, if necessary, Harry adds back to some of these trimmed pieces. Eventually we give Tom only untrimmed pieces he prefers, while Dick and Harry get only the pieces they prefer also. Then the trimmings will be dispensed using advantages Tom has.

Step 1: Tom cuts $t + d + h$ equal pieces $X_1, \cdots, X_{t+d+h}$.

Step 2: Dick trims the largest $d - 1$ pieces, if necessary, so that they are equal to the dth largest piece. This results in sets D_a of altered pieces and

D_u of untrimmed pieces that Dick views as tied for the largest in size. The remaining pieces $D^{'}$ are all strictly smaller to Dick. We have $|D_a| \leq d-1$ and $|D_a \cup D_u| \geq d$.

Step 3: Harry indicates the set H consisting of the h largest pieces in his view (after Dick's trimming). If X_j is the hth largest piece, then H contains all pieces strictly larger than X_j and possibly some pieces equal in size to X_j in Harry's opinion. Nevertheless, Harry indicates exactly h pieces even if there are many pieces equal to X_j.

If $|D_a \cup D_u \cup H| \geq d+h$, give the pieces in H to Harry, any remaining pieces in D_a to Dick, and enough pieces in D_u to Dick to total d pieces. Since $|D_a| \leq d-1$, all trimmed pieces are given to Dick or Harry. The remaining pieces going to Tom are all untrimmed. The difficulty as we proceed is to make sure that all trimmed pieces go to either Dick or Harry.

If $|D_a \cup D_u \cup H| < d+h$, we don't have enough pieces to give to Dick and Harry yet, and we refer to this case as an *insufficient stage*. When this is the case we go on to:

Step 4: Dick trims all of the pieces in $D_a \cup D_u$ so that they equal the largest piece in $D^{'}$. Harry again indicates the largest h pieces as in Step 3. In case of ties for the hth largest piece, Harry's choice of which pieces to include in H is arbitrary except for the rule that any piece in the previous set H not trimmed by Dick in this step must also be included in the new set H.

We repeat Step 4 until the first time $|D_a \cup D_u \cup H| \geq d+h$. We call this the *first sufficient stage*. This must eventually occur since D_a increases each time Step 4 is repeated.

Step 5: We may assume that there is a last insufficient stage where $|D_a \cup D_u \cup H| < d+h$ and that the next stage is sufficient. Denote by D_a^*, D_u^*, $D^{'*}$, and H^* the appropriate sets at the first sufficient stage. Note that D_a^* is the set of pieces generated by trimming pieces in $D_a \cup D_u$ and $|D_a^* \cup D_u^* \cup H^*| \geq d+h$. If $|D_a^* \cup H^*| \leq d+h$, again our task is easy. Give the pieces in H^* to Harry, the remaining pieces in D_a^* to Dick. Since we are now at a sufficient stage, there are enough still untrimmed pieces in D_u^* that Dick considers largest to give him a total of d pieces. Note that Tom is left only untrimmed pieces.

Now let us assume that $|D_a^* \cup H^*| > d+h$. We must examine the pieces in H^* in detail.

A typical piece in D_a^* appears in Figure 11.1.

E_i is the part of X_i trimmed during all but the last insufficient stage. (If $X_i \in D_u$ then $E_i = \emptyset$.)

F_i is the part of X_i trimmed as we enter the sufficient stage, and $Y_i = X_i - E_i - F_i$.

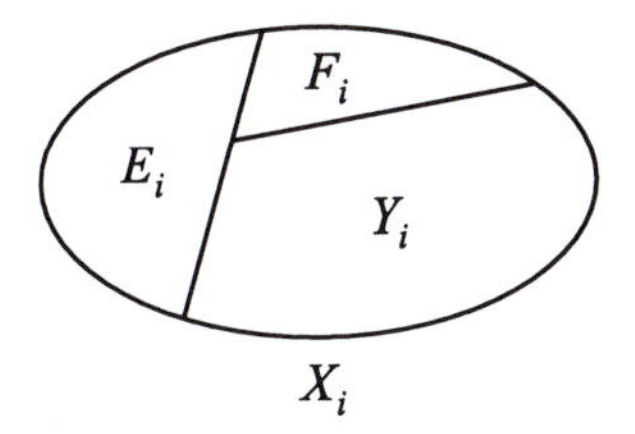

Figure 11.1.

By renumbering the pieces we may assume we have the four disjoint subsets:
$\{Y_1, \cdots, Y_r\} \equiv$ those pieces in D_a^* trimmed from pieces in H,
$\{Y_{r+1}, \cdots, Y_s\} \equiv$ those pieces in D_a^* not trimmed from pieces in H,
$\{X_{s+1}, \cdots, X_v\} = D' \cap H$,
$\{X_{v+1}, \cdots, X_{d+h}, \cdots, X_w\} = D' \cap H^* \setminus H$.

Remarks:

(1) $\{Y_1, \cdots, Y_s\} = D_a^*$.

(2) $\{X_{s+1}, \cdots, X_v\} \subset H^*$ since these pieces were in H and remain untrimmed.

(3) All pieces in H correspond to pieces found in $\{Y_1, \cdots, Y_r, X_{s+1}, \cdots, X_v\}$.

(4) We know $w = |D_a^* \cup H^*| > d + h$. We may also assume that the pieces $X_{v+1}, \cdots, X_w$ are listed in decreasing order of size as viewed by Harry.

(5) Some of the Y_i may be in H^*; in particular, the set A of all Y_i such that $\mu_h(Y_i) > \mu_h(X_{d+h})$ must be in H^* since $X_{d+h} \in H^*$. Also, by the construction $\mu_h(X_i) \geq \mu_h(X_{t+d})$ if $s + 1 \leq i \leq v$ since such $X_i \in H$ and $X_{d+h} \in H^* \setminus H$.

We will now construct exactly h pieces, which we will give to Harry, that are all among the first $d + h$ pieces and that are all at least as large as X_{d+h}. Dick will get the rest of these $d + h$ pieces, thereby guaranteeing that Tom receives only untrimmed pieces.

Step 6: First, we give Harry the pieces $A \cup \{X_{s+1}, \cdots, X_{d+h}\}$. These pieces are all in H^* and there are fewer than h of them since $w > d + h$.

Next, for all $Y_i \in D_a^* \setminus A$:

(1) If $\mu_h(Y_i \cup F_i) \leq \mu_h(X_{d+h})$, let $Z_i = Y_i \cup F_i$.

(2) If $\mu_h(Y_i \cup F_i) > \mu_h(X_{d+h})$, have Harry trim off a piece G_i from F_i such that $\mu_h(Y_i \cup G_i) = \mu_h(X_{d+h})$, and let $Z_i = Y_i \cup G_i$.

Let S be the set of all such pieces $Z_i = Y_i \cup G_i$. Note that if $Y_i \cup F_i \in H$, then $\mu_h(Z_i) = \mu_h(X_{t+d})$ since $X_{t+d} \notin H$.

Among the first $d+h$ pieces, Harry has already been given all of the pieces he considers larger than X_{d+h} and some pieces equal in size to X_{d+h}. Now let Dick give Harry enough pieces in S with $\mu_h(Z_i) = \mu_h(X_{t+d})$ to give him a total of h pieces. There must be enough such pieces since every piece in $H \setminus \{X_{s+1}, \cdots, X_v\}$ corresponds to a piece in A or S, and, for those not in A, $\mu_h(Z_i) = \mu_h(X_{t+d})$ as noted above. Harry has now received h pieces he views as largest.

Dick gets all of the pieces Z_i that he didn't give to Harry. There are exactly d such pieces, and to Dick each is as large as any X_i with $i > s$. Since Dick chooses which Z_i to give away, Harry receives no piece larger than Dick's smallest. Tom gets the remaining h untrimmed pieces.

Step 7: All that remains is to divide the excess pieces which are of the forms E_i, $E_i \cup F_i$, or $E_i \cup F_i \setminus G_i$. Let K_i denote any of these forms. Either Dick or Harry received $X_i \setminus K_i$, so suppose it was Dick (the two cases are symmetric). Have Harry cut K_i into $t + d + h$ equal pieces. Dick chooses d of them, and then Tom chooses t of them, and Harry gets the rest. The argument for over-all envy-freeness is similar to the 2 : 2 : 1 example discussed earlier.

It is easy to see that this method uses a bounded number of cuts, depending only on the values t, d, and h. If we let $n = t + d + h$, then we can count the number of cuts and find that fewer than $\frac{5}{6}n^2$ cuts always suffice. (See Exercise 11.2.)

11.5 EXERCISES

11.1. How can envy-free division into unequal portions be accomplished for two players using a bounded algorithm?

11.2. Verify the bound on the number of cuts in Section 11.4.

11.3. The Cut Near-Halves Algorithm may require as many as $\lceil \log_2(a + b) \rceil$ cuts for dividing in the ratio $a : b$. We saw some examples in Section 3.3 where fewer cuts sufficed. Also, this algorithm is restricted in that, once a piece is chosen, that piece is never cut again. Perhaps we can do better than $\lceil \log_2(a + b) \rceil$ by using a more general finite algorithm as defined in Section 2.1. (Of course, the Moving Knives Exact Division Algorithm in Section 5.4 requires only 2 cuts, but this is a continuous algorithm, not a finite one.)

Show that for the particular ratio of $1 : n$ *any* finite algorithm requires at least $\lceil \log_2(n + 1) \rceil$ cuts.

11.6 PROJECT

11.1. How could Ramsey partitions into three or more subsets be defined? Can they be used in fair division methods?

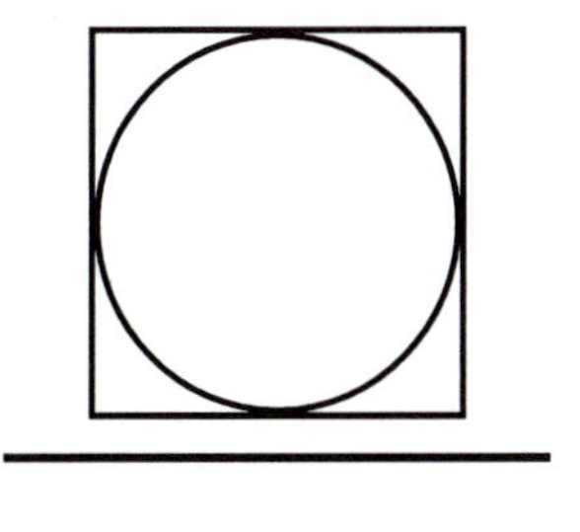

Solutions

Chapter 1

1.1. In order to guarantee himself a fair share the cutter must cut what he considers halves, so can do no better than receive 1/2. If the chooser disagrees with that cut, she will get more than a half (by her estimation). So you prefer to be the chooser.

1.2. If Dick does not view Tom's cut as exactly $1/3, 2/3$, there will be at most two pieces Dick views as at least 1/3 after his cut. (Consider the two cases where Dick views X_1 as bigger or smaller than 1/3.) Choosing last, he may be forced to take an unacceptable piece.

1.3. Harry has a chance that the piece offered to him may be what he considers more than 1/3. If Dick exercises his option to trim and cuts away a piece Tom thinks has positive value, then Tom will eventually get strictly more than 1/3. If Tom cuts what Dick considers less than 1/3, then Dick will get more than 1/3. So all have the possibility of getting more than 1/3.

1.4. (a) Tom gets exactly 2/3 of exactly 1/2, or exactly 1/3, by his estimation. Dick gets exactly 2/3 of *at least* 1/2, and so he may get more than 1/3 (when he disagrees with Tom's first cut). Harry gets two chances to get strictly more than 1/3, which will happen if either of the other two does not cut what Harry considers exact thirds in the second round of cuts.

(b) Harry could cut each of the pieces held by Tom and Dick into exact thirds and let each choose two of the three pieces. Harry would then get an exact third of the cake, and hence he will prefer to use the original algorithm.

1.5. For three players, the Trimming Algorithm may require three cuts (when Dick trims). Note that Cut and Choose can be accomplished with just one cut, even when the portion to be divided is in two pieces. (See One-Cut Suffices in Section 2.2.) For Successive Pairs three persons use five cuts.

For four players, the Trimming Algorithm can require three cuts before the first piece is chosen, two more cuts before the second piece is chosen, and one additional cut for the final Cut and Choose. Thus six total cuts may be required. For Successive Pairs it will require five cuts for three players to each get 1/3 and each of the three must use three additional cuts in order to cut their portions in exact fourths. Thus 14 total cuts are required. (See Section 2.3.)

1.6. (a) Only Will. Both Phil and Jill may cut a piece the other considers very large and Will may choose those pieces.

(b) Jill can get all of one of her original thirds so she is guaranteed 1/3. Phil is guaranteed at most 1/5 (in the case five pieces are considered equal fifths and the other worthless, see Exercise 1.10). If Will views Jill's first cuts as (roughly) $1/2, 1/2, 0$, he could wind up with as little as 1/4.

(c) Phil is guaranteed at most 1/5. Will is not guaranteed any positive amount. He may consider Jill's original cut (roughly) 1, 0, 0 and not get any from the large piece. However Jill is guaranteed 1/3. She will get at least what she considers the second largest of the six pieces on her first choice. After that choice at least three of the remaining four pieces are sufficiently large to combine with her first choice to give the 1/3. She is guaranteed to get one of those three.

1.7. Many could be devised. For example, Jill cuts four pieces and they are chosen in the order PWWJ. Phil may not think any piece is worth 1/3; Will may think Phil gets more than 2/3 on Phil's first choice; Jill can't cut four pieces each worth 1/3.

1.8. (a) Tom get X_3, Dick gets X_1, and Harry gets X_2.

(b) Tom cuts equal thirds; Dick must think some piece has value at least 1/3, as must Harry.

(c) Give Tom either X_2 or X_3, and let Dick and Harry play Cut and Choose on the other two pieces.

(d) Give either X_2, X_3, or X_4 to Tom, and let the other three play the three person game on the other three pieces.

(e) From part (d), if only Tom likes a particular piece, give it to him and let the others play the three person game on the rest of the cake. If only two players like a particular piece, say Tom and Dick, give that piece to Dick and let the other three play the three person game on the other three pieces. Thus we may assume each column has at least three ones in it, so we may assume without loss of generality that the matrix takes the form (no assumption is made on entries left blank):

	X_1	X_2	X_3	X_4
Tom	1	1	1	1
Dick	1	1		
Harry	1	1		
Amy			x	y

If either x or y is one, say $x = 1$, give X_3 to Amy, X_4 to Tom, and let Dick and Harry divide $X_1 \cup X_2$.

Finally, suppose $x = y = 0$. Then we have (since there are three ones in each column):

$$\begin{matrix} 1 & 1 & 1 & 1 \\ 1 & 1 & 1 & 1 \\ 1 & 1 & 1 & 1 \\ z & w & 0 & 0 \end{matrix}$$

Either $z = 1$ or $w = 1$. If $z = 1$, give X_1 to Amy and distribute the others arbitrarily.

1.9. Only Tom. The division of E may not be envy-free.

1.10. (a) Yes. Choose in the order $P_1, P_2, \cdots, P_n$.

(b) For $n = 3, (n^2 + n)/2 = 6$. All three players could consider X_1 and X_2 acceptable and X_3 unacceptable.

(c) You could have as many as $n(n-1)$ "yes" responses and not have a fair division possible. The matrix in Exercise 1.8 could have its only zeros in a single column. (We aren't assuming all ones in any row.)

We prove by induction that $n(n-1)+1$ "yes" responses always allows a fair assignment of the pieces. For $n = 1, 2$ the result is clear. For $n = 3$, we assume the 3×3 matrix has at least seven ones. If there are two zeros, assign an acceptable piece to a player so that the coresponding row and column contains both zeros. The reduced 2×2 game has all ones as entries.

Now assume $n \geq 3$ and there are $n-1$ zeros in the matrix. (Fewer zeros makes the task easier.) Assign the first acceptable piece so that the corresponding row and column contains at least two zeros. This is always possible. If two zeros both lie in a common row or column, go to a one guaranteed to be in that row or column, and assign the corresponding piece. Assume two zeros are in different rows and columns, say $a_{ij} = 0, a_{kl} = 0$. If $a_{il} = 1$ or $a_{jk} = 1$ make the corresponding assignment. If $a_{il} = 0$ we have two zeros in the same row and we are back to the first case. In the reduced $(n-1)\times(n-1)$ matrix there are at least $(n^2 - n + 1) - (n + n - 1 - 2) = n^2 - 3n + 4$ ones left. By induction $(n-1)(n-2) + 1 = n^2 - 3n + 3$ ones suffice for a fair division assignment on the reduced $n - 1$ player game.

1.11. (a) At least 1/6 on your first choice. No amount can be guaranteed on the last choice.

(b) Order the pieces in value $a_1 \geq a_2 \geq a_3 \geq a_4 \geq a_5 \geq a_6$ and assume $a_6 = (1/6) - \epsilon$. Then $a_1 \geq (\frac{1}{5})[(1 - (1/6 - \epsilon)] = 1/6 + \epsilon/5$. So $a_1 + a_6 \geq 1/3 - (4/5)\epsilon$. For $0 \leq \epsilon \leq 1/6$, this expression takes on values between 1/3 (when $\epsilon = 0$ and all pieces are considered of equal size) and 1/5 (when $\epsilon = 1/6$ and the last piece is considered worthless). Note that the worst case is when the six pieces have values 1/5, 1/5, 1/5, 1/5, 1/5, 0.

Chapter 2

2.1. (a) Three cuts may be needed. If Dick and Harry accept different pieces, only Tom's original two cuts are needed. Only 1.8 (c) is troublesome. Here we can give X_3 to Tom and let Dick and Harry divide $X_1 \cup X_2$ using one more cut.

(b) Six cuts may be needed: three to cut in quarters and three more for the three-player reduced problem.

(c) $4 + 6 = 10$ cuts.

(d) The number of cuts for n players can be as large as $1 + 2 + \cdots + (n-1) = n(n-1)/2$, which is the same as the Trimming Algorithm.

2.2. $D^*(4) = 6, D^*(5) = 8, D^*(6) = 11, D^*(7) = 15, D^*(8) = 18$. The case $n = 4$ is worse, which leads to a worse estimate when $n = 7$ (even though both algorithms start with a $4 : 3$ division). The cases $n = 5$ and 6 are the same for both algorithms, but $n = 8$ is worse.

2.4. P_1 cuts $A \cup B$, each worth 1/2. We may assume P_2 and P_3 view A as at least 1/2.

(i) If $\mu_4(B) \geq 1/3$, then P_1 and P_4 share B while P_2 and P_3 share A, each using only one more cut.

(ii) Next assume $\mu_4(A) > 2/3$. If either $\mu_2(B) \geq 1/3$ or $\mu_3(B) \geq 1/3$, proceed as in case (i).

(iii) If $\mu_i(A) > 2/3$ for $2 \leq i \leq 4$, P_2, P_3, and P_4 can share A with two cuts using the algorithm of Section 2.5. Each gets at least $\frac{1}{4} \cdot \frac{2}{3} = \frac{1}{6}$.

2.5. Although it seems unlikely, it has not yet been shown impossible.

2.6. With $D(1) = 0$ and $D(2t) = 2t - 1 + 2D(t)$, $D(2t+1) = 2t + D(t) + D(t+1)$, $t \geq 1$, we wish to show $D(n) = nk - 2^k + 1$ for $n \geq 1$, where $k = \lceil \log_2 n \rceil$. Proceeding by induction, for $n = 1$, $nk - 2^k + 1 = 0 - 1 + 1 = 0$ checks. For $n = 2$, $D(2) = 1 + 2D(0) = 1 = nk - 2^k + 1 = 2 - 2 + 1$. Assuming the result for $1, 2, \cdots, n-1$, let $n > 2$. Write $n = 2^{k-1} + r$, where $1 \leq r \leq 2^{k-1}$ so that $\lceil \log_2 \ n \rceil = k$.

Case 1: If $r = 2s$, $1 \leq s \leq 2^{k-2}$ then $D(n) = (2^{k-1} + 2s - 1) + 2D(2^{k-2} + s) = (2^{k-1} + 2s - 1) + 2((2^{k-2} + s)\lceil \log_2(2^{k-2} + s) \rceil - 2^{\lceil \log_2(2^{k-2}+s) \rceil} + 1) = (2^{k-1} + 2s - 1) + 2((2^{k-2} + s)(k-1) - 2^{k-1} + 1) = (k-2)2^{k-1} + 2ks + 1 = (2^{k-1} + 2s)k - 2^k + 1 = nk - 2^k + 1$ as required.

Case 2: If $r = 2s + 1$, $0 \leq s \leq 2^{k-2} - 1$, then, if $s \neq 0$, $D(n) = (2^{k-1} + 2s) + D(2^{k-2} + s + 1) + D(2^{k-2} + s) = (2^{k-1} + 2s) + ((2^{k-2} + s + 1)\lceil \log_2(2^{k-2} + s + 1) \rceil - 2^{\lceil \log_2(2^{k-2}+s+1) \rceil} + 1) + ((2^{k-2} + s)\lceil \log_2(2^{k-2} + s) \rceil - 2^{\lceil \log_2(2^{k-2}+s) \rceil} + 1) = (2^{k-1} + 2s) + (2^{k-2} + s + 1)(k-1) - 2^{k-1} + 1) + ((2^{k-2} + s)(k-1) - 2^{k-1} + 1) = (2^{k-1} + 2s + 1)k - 2^k + 1 = nk - 2^k + 1$ as required. If $s = 0$, $\lceil \log_2(2^{k-2} + s) \rceil = k - 2$ in the above expression, in which case we get $D(n) = (2^{k-1} + 1)k - 2^k + 1 = nk - 2^k + 1$ again.

Chapter 3

3.1. (a) 10, 10, 3, 3, 3, 1, 1, 1, 1

(b) 10, 10, 1, $\cdots$, 1 (11 ones)

(c) 1, $\cdots$, 1 (22 ones)

3.2. The longest path in the Cut Near-Halves tree is $33 \to 16$, $\mathbf{17} \to 8$, $\mathbf{9} \to 4$, $\mathbf{5} \to 2$, $\mathbf{3} \to 1$, $\mathbf{2} \to 1$, 1, which requires 6 cuts. Ramsey uses 8 cuts. The numbers in bold correspond to the pieces chosen by the noncutter. Note $2^5 < 33 \leq 2^6$ so 6 cuts may be required.

3.3. $1:2:7$

Ramsey
$$\left.\begin{array}{ll} 1:2 \longrightarrow 1,1,1 & \text{2 cuts} \\ 3:7 \longrightarrow 3,3,1,1,1,1 & \text{5 cuts (twice)} \end{array}\right\} \text{12 cuts total}$$

Cut Ones
$$\left.\begin{array}{ll} 1:2 \longrightarrow 1,1,1 & \text{2 cuts} \\ 3:7 \longrightarrow 1,\cdots,1 & \text{9 cuts (twice)} \end{array}\right\} \text{20 cuts total}$$

3.4. $2:3:5$
$$\left.\begin{array}{ll} 2:3 \longrightarrow 2,1,1,1 & \text{3 cuts} \\ 5:5 = 1:1 & \text{1 cut (twice)} \end{array}\right\} \text{5 cuts total}$$

$5:3:2$
$$\left.\begin{array}{ll} 5:3 \longrightarrow 3,2,1,1,1 & \text{4 cuts} \\ 8:2 = 4:1 \longrightarrow 1,1,1,1,1 & \text{4 cuts (twice)} \end{array}\right\} \text{12 cuts total}$$

3.5. $2:3:4$
$$\left.\begin{array}{ll} 2:3 \longrightarrow 2,1,1,1 & \text{3 cuts} \\ 5:4 \longrightarrow 4,1,1,1,1,1 & \text{5 cuts (twice)} \end{array}\right\} \text{13 cuts total}$$

The order $2:4:3$ uses only 6 cuts but $3:4:2$ uses 14 cuts.

3.6. Player P_2 will make the first cut. If P_1 views the pieces as equal, she will receive only 1/2 of the cake. Hence P_2 must cut again and P_1 chooses. Again P_1 may view the pieces as equal and hence get only 1/4 more. With 2 cuts by P_2, P_1 may get only 3/4 of the cake, but needs 4/5. So another cut is required.

3.7. The ratio $7:2$ requires as many as 4 cuts using Cut Near-Halves. However, if the first cut is in the ratio $6:3$, then no matter which piece Beth chooses, the other piece must be divided in the ratio $2:1$, which requires only 2 more cuts.

3.8. Consider the ratio $88:65$, for example. Using Cut Near-Halves as many as 8 cuts may be needed. But suppose Ann cuts $10:143$. Depending on Beth's choice, the piece of size 10 goes to Ann or Beth, leaving ratios of $78:65 = 6:5$ or $88:55 = 8:5$ respectively. In either case, at most 4 more cuts suffice for a total of only 5 cuts.

Chapter 4

4.1. (a) With the parallel cuts of all players on one copy, starting at the left edge, move to the right until the first cut is encountered. Give the piece to the left of that cut to the person who made the cut. (Ties can be broken arbitrarily). Erase all cuts made by the person who received the piece; erase all other left-most (first) cuts for all players. Repeat the process on the unassigned portion until all n players receive a piece.

(b) The example shown in Figure 4.2 has the property.

4.2. (a) No, if he wins the smaller bet and loses the larger, he loses. In all other cases he wins.

(b) Considering the tails of the bets to the extreme right and extreme left of the figure, if there is an odd number of equal bets, on one tail Tom will win fewer bets than he will lose, so the wager is not risk-free. Note that for a particular friend, if Tom wins on one tail he loses on the other.

For four friends (or any even number) Tom can simply break the group of friends into groups of two each and he can then make bets with each group of two as before. In this case risk-free bets are possible.

For unequal bets the situation is more complicated. For example, suppose Tom makes a \$2 bet with Dick and a \$1 bet with each of Harry and Amy. If the cuts are as follows (showing Tom's payoff):

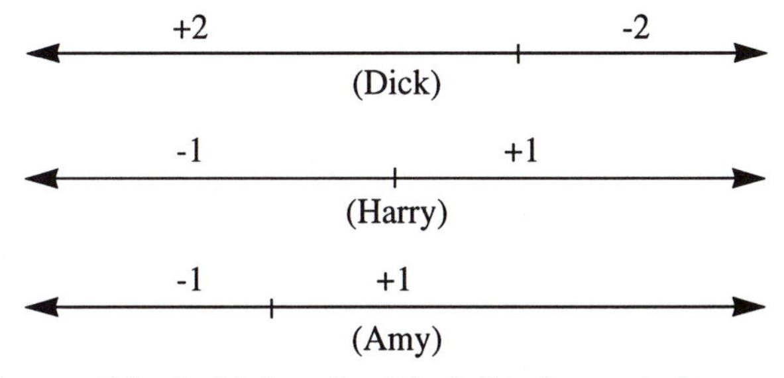

then the overall bet is risk-free. But it isn't if he has made the opposite bet with each player.

Similar examples are easily constructed for any number of players when unequal bets are allowed.

4.3. (a) The car is given to Abe; he gets a third of its value as his fair share but pays the estate $2/3 \times \$15,100 = \$10,066.67$. From that amount Jill receives $1/3 \times \$14,300 = \$4,766.67$ and Mary gets $1/3 \times \$13,200 = \$4,400$. So the cash excess generated by the car is $\$10,066.67 - \$4,766.67 - \$4,400 = \$900$. Similarly the other excesses generated are:

Boat: $2/3(\$8,200) - 1/3(\$7,300 + \$7,300) = \$600$

Piano: $2/3(\$3,000) - 1/3(\$2,500 + \$2,900) = \$200$

(b) $2/3a - 1/3(b + c)$

(c) With $a \geq b \geq c$, the expression in (b) can only be nonpositive if $a = b = c$. So any item on which there is any disagreement will generate cash excess.

4.4.

	Jill	Abe	Mary	
Car	14,300	15,100	13,200	
Boat	8,200	7,300	7,300	
Piano	3,000	2,500	2,900	
Value of the Estate	25,500	24,900	23,400	
Fair Share of Estate	12,750	7,470	4,680	
Items Received	Boat, Piano	Car	—	
Cash Adjustment	+$1,550	-$7,630	+$4,680	(Cash excess $1,400)
Cash Bonus Beyond Fair Share	$700	$420	$280	(If distributed 50:30:20)

If Jill bids high the cash excess is $.5j - .3a - .2m$. If Mary bids high the cash excess is $.8m - .5j - .3a$. Assume in the first case Jill bids $j = x > y = m$ while in the second case $m = x > y = j$. Assume Abe's bid does not change. Then the difference in cash excess is $(.5x - .2y) - (.8x - .3y) = .1y - .3x < 0$. So more excess is generated when Mary's bid is high. Note, in any case, that the cash excess generated by any item is non-negative.

4.5. (a) The algorithm results in P_1 getting A and no money while P_2 gets B plus \$3. Thus, P_1 thinks she received \$9 while P_2 received \$11.

(b) If $a_{ij} = P_i$'s opinion of P_j's share and $A = [a_{ij}]$, then

$$A = \frac{1}{3}\begin{bmatrix} 31 & 28 & -5 \\ 28 & 31 & 22 \\ 19 & 19 & 34 \end{bmatrix}$$

and the division is envy-free.

(c) Suppose P_i bids a_i on a single item, with $a_1 \geq a_2 \geq \cdots \geq a_n$. Then the adjusted bids will be $a_1, b, b, \cdots, b$, with $a_1 \geq b \geq a_2$. Player P_1 will first receive the item and return $((n-1)/n)a_1$ to the estate. Each of $P_2, \cdots, P_n$ receive b/n in cash. Then all n players receive $e = (a - b)(n-1)/n^2$ in cash excess. Player P_1 thinks he or she has received $a_1/n + e$ while $P_i, i \neq 1$, has received $b/n + e$. For $i \neq 1$, player P_i thinks P_j has received $b/n + e, 2 \leq j \leq n$, while P_1 has received $e + a_i - a_1(n-1)/n$ with $a_1 \geq b \geq a_i$. Thus the division is envy-free.

4.6. For special regular triangulations, label the vertices and points along the edges as shown. At each such point the labeled room is free. Now assign each vertex in the triangulation to one of the renters in such a way that the vertices of each small triangle are assigned to a different person. At each vertex in the interior with coordinates (a, b, c) have the person assigned that vertex label it A (or B, or C) if room A (or room B, or room C) is preferred at cost a (or b, or c respectively).

According to Sperner's Lemma, some small triangle will have its vertices labeled using all three letters. Further refinements provide the rent distribution point (a, b, c) required. Note that this envy-free distribution can't have a zero entry since any person paying rent will envy a person paying no rent. Thus the point found is in the interior of the triangle.

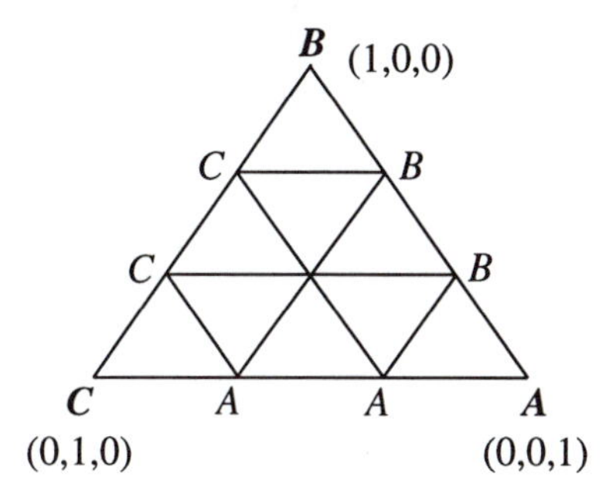

Chapter 5

5.1. Add $a \to c$ and $b \to d$ to Figure 5.1.

5.2. We are given that $\mu_i(X_i) = (1 + \epsilon_i)/n, \epsilon_i > 0$ for $i = 1, 2, \cdots, n$. Find k_i so that $[(nk_i - 1)/((n+1)k_i - 1)]\mu_i(X_i) > 1/(n+1)$. With k_i so chosen, P_i cuts X_i into $(n+1)k_i - 1$ equal portions and lets P_{n+1} choose k_i of them. Then P_{n+1} gets $[k_i/((n+1)k_i - 1)]\mu_{n+1}(X_i) > (1/(n+1))\mu_{n+1}(X_i)$ from P_i. Hence P_{n+1} gets more than $1/(n+1)$ of the entire cake. On the other hand, P_i gets $[(nk_i - 1)/((n+1)k_i - 1)]\mu_i(X_i) > 1/(n+1)$ by the choice of k_i.

5.3. Examine cases depending on who shouts "cut."

Case 1: Player with the left-most knife shouts "cut." The shouter is satisfied with the left piece; the middle cutter gets the middle piece, which is equal to the right piece and at least as large as the left piece, since the middle cutter did not shout "cut." The player with the right-most knife thinks the right piece is at least as large as the middle piece and no smaller than the left piece, since he did not shout "cut."

Case 2: Player with the middle knife shouts "cut." The player with the left-most knife gets the middle piece, which is considered at least as large as the right piece and no smaller than the left (since the player didn't shout "cut"). The person with the right knife gets the right piece and is similarly satisfied.

Case 3: Player with the right-most knife shouts "cut." The shouter gets the left piece, the player with the left-most knife gets the middle piece, and the middle cutter gets the right piece. That all are satisfied follows as above.

5.4. First, Dick must agree with Tom before a rotation of 180° of the directed knife. If Dick, for example, thinks a larger piece lies to the right of the directed knife for $\theta = 0°$, then at $\theta = 180°$ when the knife will be in the same position but directed in the other direction the larger piece will be to the left of the knife. Under reasonable continuity assumptions, somewhere in between Dick will think the cake is bisected. Finally, since neither Dick nor Harry yelled "cut," they will consider their pieces at least as big as Tom's. (After a 180° rotation a curved knife might not coincide with the original position.)

5.5. (a) We must have $b - a = 1/3$ in order to have $\mu_F(A_{ab}) = 1/3$. Also $\mu_G(A_{ab}) = 1/3$ when $\int_a^b (-2x + 2)dx = 1/3 = a^2 - b^2 + 2b - 2a$. Substituting $b = a + 1/3$ gives $a = 1/3, b = 2/3$.

(b) If $0 < \alpha < 1$ then $\mu_F(A_{\alpha b}) = \alpha$ whenever $b - a = \alpha$. Also $\mu_G(A_{0\alpha}) > \alpha$ while $\mu_G(A_{1-\alpha,1}) < \alpha$. Hence, by the Intermediate Value Theorem, for some position of the knives between 0 and $1 - \alpha$ the desired cut can be made.

(c) Note that $\mu_F(A_{1/2-a,1/2+a}) = \mu_G(A_{1/2-a,1/2+a})$ for any $a \epsilon [0, 1/2]$. So we need $\alpha = \int_{1/2-a}^{1/2+a} (-2x + 2)dx$. Given $\alpha \epsilon [0, 1]$ solve this equation for a to get $a = \alpha/2$.

5.6. (a) Since $\mu_F(A_{ab}) = 3/4, b - a = 3/4$ and $0 \leq a \leq 1/4$. But $12 \int_a^{a+3/4} (x - 1/2)^2 dx = 4[\left(a + \frac{1}{4}\right)^3 + \left(\frac{1}{2} - a\right)^3] < 3/4$ if $0 \leq a \leq 1/4$.

(b) Now $b - a = 1/4$ and $12 \int_a^{a+1/4} (x - 1/2)^2 dx = 4[(a - 1/4)^3 + (1/2 - a)^3] = 3a^2 - \frac{9}{4}a + 7/16$ and this is 1/4 for two values of $a \epsilon [0, 3/4]$.

5.7. (See the result in Section 6.1 for the general case.)

Case 1: If there is a player, say Tom, who accepts only one piece, give that piece to him. Dick and Harry must find different pieces of the remaining two acceptable, since fair assignments are possible. If one of them likes only one of the remaining pieces, let him have it. If both accept either, let either choose first.

Case 2: If every player accepts at least two pieces, there are two possibilities. (a) If some two players, say Tom and Dick, accept only the same two pieces, choose in the order T, D, H (or D, T, H). (b) Otherwise, let Tom choose first. If either Dick or Harry finds only one of the remaining pieces acceptable, let him have it. The other will accept the remaining piece. If both Dick and Harry accept both remaining pieces, let them choose in either order.

5.8. (i) *Trimming Algorithm.* The only difference from the simple fair division algorithm is that each person after the first cutter will increase (rather than decrease) the size of the portion to size $1/n$. The same number of cuts is used.

(ii) *Successive Pairs Algorithm.* The only change required is that players will choose smallest rather than largest pieces. The same number of cuts is used.

(iii) *Kuhn Algorithm.* Portions are acceptable if they have size no more than $1/n$. Nothing else changes. The same number of cuts is used.

(iv) *Cut Near-Halves Algorithm.* We assume two persons are to do a chore in some unequal rational ratio. The person with the smaller remaining assignment cuts near-halves, and the other chooses the portion he or she prefers to accept at the cut values. Ratios are reduced as before. The same tree, hence the same number of cuts, results.

5.9. Yes. Give the third cutter a portion smaller than his or her cut but larger than the portion cut by the second cutter. Both the first and second cutter feel more than $1/3$ has been claimed, so playing Cut and Choose on the remainder will give each strictly less than $1/3$.

5.10. The knife is moved from left to right and the first player to think the right piece has been reduced to 1/3 shouts "cut" and receives that piece. Nothing else in the algorithm changes.

5.11. If P_1 shouts "cut," P_1 gets X_1, P_2 gets $X_2^{''}$, and P_3 gets $X_2^{'}$.
If P_2 shouts "cut," P_2 gets X_1, P_1 gets $X_2^{''}$, and P_3 gets $X_2^{'}$.
If P_3 shouts "cut," P_3 gets X_1, P_1 gets $X_2^{''}$, and P_2 gets $X_2^{'}$.

5.12. They run the risk of another player shouting cut and thus that player receiving a portion smaller than what the "cheater" considers $1/3$. Depending on the positions of the other knives, the "cheater" may be left with a portion he or she considers bigger than 1/3. In certain cases it is clear some strategies are risk-free for a while.

5.13. The only change needed is, when assigning portions, go the position of the *farthest right* (rather than farthest left) cut on any given assignment. Proceed by induction as in the original algorithm.

5.14. Many can be given, for example, $f(x, y) = 2y$ and $g(x, y) = -2y + 2$.

Chapter 6

6.1. (a) In Condition 1 if $v_{ii} < v_{ij}$ and $v_{jj} < v_{ji}$ then $v_{ii} + v_{jj} < v_{ij} + v_{ji}$ so switching increases S. Any switch in Condition 2 increases S by definition.

(b) Since all of the row sums of V are one, the average value of S over all permutations is one. Hence, $S \geq 1$ for at least one permutation.

6.2.

$$\begin{bmatrix} \mathbf{.35} & .4 & .25 \\ .25 & \mathbf{.35} & .4 \\ .4 & .25 & \mathbf{.35} \end{bmatrix}$$

If any pair switches pieces, one player will get less, but using a 3-player permutation everyone could get a piece worth .45, which is better for everyone.

6.3. Assignment $\begin{bmatrix} \mathbf{.5} & .4 & .1 \\ .3 & \mathbf{.3} & .4 \\ .45 & .25 & \mathbf{.3} \end{bmatrix}$ satisfies Condition 5, but assignment

$\begin{bmatrix} .5 & \mathbf{.4} & .1 \\ .3 & .3 & \mathbf{.4} \\ \mathbf{.45} & .25 & .3 \end{bmatrix}$ satisfies Condition 3.

6.4. (a) Using the matrix V in the previous answer would yield the assignment shown below with $S = 1.15$. However, the second assignment shown in Solution 6.3 above gives $S = 1.25$.

$$\begin{bmatrix} \mathbf{.5} & .1 & .4 \\ .3 & \mathbf{.4} & .3 \\ .45 & .3 & \mathbf{.25} \end{bmatrix}$$

(b) It is convenient to prove the stronger result that the greedy algorithm produces a value $S \geq 1$ whenever all of the row sums are at least one. Use induction on the size of V. The claim is true for $n = 1$.

If v_{11} is maximal, delete the first row and column to obtain the matrix V'. Since $v_{11} \geq v_{j1}$ for all $j > 1$, all row sums of $\frac{1}{1-v_{11}}V'$ are at least one and so by induction the sum S' obtained by the algorithm from V' is at least one. Moreover, the assigned pieces from V' are the same as those from V.

Thus, $S' = \left(\frac{1}{1-v_{11}}\right)(v_{22} + \cdots + v_{nn}) \geq 1$ which implies $S = v_{11} + v_{22} + \cdots + v_{nn} \geq 1$.

6.5. Since there are only finitely many permutations of the columns, there must be at least one that maximizes S. By Exercise 6.1 this must also satisfy 1 and 2.

6.6. (a) If $e_{ij} = v_{ii} - v_{ij} \geq 0$ for all i and j, then each player P_i prefers his present piece X_i to every other piece X_j.

(b) If $V = \begin{bmatrix} .4 & .1 & .5 \\ .1 & .4 & .5 \\ .1 & .1 & .8 \end{bmatrix}$ then $E = \begin{bmatrix} 0 & .3 & -.1 \\ .3 & 0 & -.1 \\ .7 & .7 & 0 \end{bmatrix}$ and

$$D = \begin{bmatrix} 0 & .3 & .3 \\ .3 & 0 & .3 \\ .3 & .3 & 0 \end{bmatrix}.$$

(c) In a permutation where $D \geq 0$, a largest element in each column is on the main diagonal. Any other assignment corresponds to using some elements off the main diagonal. Each such change can only decrease S (or leave it unchanged in case of ties).

However, the condition $D \geq 0$ need not be satisfied:

If $V = \begin{bmatrix} .4 & .25 & .35 \\ .1 & .4 & .5 \\ .35 & .25 & .4 \end{bmatrix}$ then $D = \begin{bmatrix} 0 & .15 & .05 \\ .3 & 0 & -.1 \\ .05 & .15 & 0 \end{bmatrix}$.
It is easily checked that $S = 1.2$ is maximal.

(d) $d_{j1} + \cdots + d_{jn} = v_{11} - v_{j1} + \cdots + v_{nn} - v_{jn} = Tr(V) - 1$ since all row sums in V are one.

6.7 (a) The assignment $\begin{bmatrix} \mathbf{.34} & .36 & .3 \\ .36 & \mathbf{.34} & .3 \\ .33 & .32 & \mathbf{.35} \end{bmatrix}$ is simple fair but does not satisfy any of the conditions.

(b) If an assignment is envy-free, then $E \geq 0$ and S is maximized as in Exercise 6.6 (c). (Simply change column to row in the proof.) Thus Conditions 1–3 are satisfied. Obviously Condition 5 is satisfied since every player gets a favorite piece. Similarly Condition 4 is satisfied, since if $\min\{v_{ii}\}$ were not maximal, some player could get a larger piece and the assignment would not be envy-free.

(c) Since the row sums of D all equal $Tr\ (V) - 1$, we must maximize the trace of V, or equivalently maximize S. In particular, by part (b), an envy-free assignment maximizes S.

6.8. Suppose all of the v_{ij} are distinct and the pieces X_i are numbered so that Condition 4 assigns X_i to P_i. We may assume that $v_{11} > v_{22} > \cdots > v_{nn}$. Any assignment where P_n does not get X_n would then give some P_i a less preferred piece. Once P_n gets X_n, the same argument shows P_{n-1} must get X_{n-1}, etc. The assignment is thus Pareto optimal.

Suppose the assignment is given by

$$\begin{bmatrix} \mathbf{.5} & .4 & .1 \\ .2 & \mathbf{.4} & .4 \\ .1 & .5 & \mathbf{.4} \end{bmatrix}.$$

Clearly the assignment of X_i to P_i is not Pareto optimal but does satisfy Condition 4.

6.9. Suppose all of v_{ij} are distinct, and $v_{11} > v_{22} > \cdots > v_{nn}$. Clearly P_1 cannot be switched from X_1 since by Condition 5 v_{11} is maximal among all entries. Once P_1 is given piece 1, P_2 cannot be switched from piece 2 by the same argument, etc. Hence the assignment is Pareto optimal.

Suppose

$$V = \begin{bmatrix} \mathbf{.5} & .5 & 0 \\ .45 & \mathbf{.4} & .15 \\ .3 & .3 & \mathbf{.4} \end{bmatrix}.$$

Clearly the assignment of piece i to player i is not Pareto optimal, although it does satisfy Condition 5.

6.10. Begin with a matching M in the value graph. Let edge e_{ij} correspond to element v_{ij} in V. Pick a minimum v_{ij} and if $e_{ij} \notin M$, delete e_{ij} from the graph and continue. If $e_{ij} \in M$, delete it and then try to extend $M - e_{ij}$ to a total matching. (Standard matching algorithms make this easy.) If no such total matching exists, the current v_{ij} is the desired value. If the total matching does exist, simply continue as above.

Chapter 8

Project 8.1. Assume we are to divide exactly in the ratio $\alpha : 1 - \alpha$. Using the notation of Theorem 8.3, proceeding again by induction and noting the division cannot be accomplished by one cut, we may suppose that at Step k we have pieces $A_{1,k} \cdots A_{k,k}$ with $a_{ij} = \mu_1(A_{i,j})$ and $b_{ij} = \mu_2(A_{i,j})$.

Furthermore, piece $A_{k,k}$ is cut at Step $k+1$ into $A_{k,k+1}$ and $A_{k+1,k+1}$ so that $a_{i,k} = a_{i,k+1}$ and $b_{i,k} = b_{i,k+1}$ for $1 \leq i \leq k-1$, where $b_{k,k+1}$ and $b_{k+1,k+1}$ are specified (non-zero) but $a_{k,k+1}$ and $a_{k+1,k+1}$ can have any consistent values.

We must find a set S such that

$$\sum_{i \in S} a_{i,k+1} = \sum_{i \in S} b_{i,k+1} = \alpha \, .$$

If $k \in S$ and $k+1 \in S$ (or neither is in S) then we contradict the nonexistence of such a set at Step k. Hence, without loss of generality, there exists $S' \subseteq \{1, \cdots, k-1\}$ such that $\alpha = \sum_{i \in S'} a_{i,k+1} + a_{k,k+1} = \sum_{i \in S'} a_{i,k} + a_{k,k+1}$.

But $a_{k,k+1}$ may have a value making this impossible for either of the following reasons:

1. Suppose $a_{k,k+1}$ is linearly independent of $\{\alpha, a_{1,k}, \cdots, a_{k-1,k}\}$ over Q.
2. There are at most 2^{k-1} subsets S', hence only 2^{k-1} values of $\alpha - \sum_{i \in S'} a_{i,k}$, and $a_{k,k+1}$ may not be one of these values.

Chapter 9

9.1. Use the same notation and assumptions as were used in the general case for n players and n cuts. Let P_5 cut equal pieces A and B. Then

(i) If $\mu_4(B) \leq 2/6$, let $P_1, \cdots, P_4$ share A using 4 cuts, and let P_5 have B.

(ii) If $\mu_4(B) > 2/6$ and $\mu_3(B) \leq 1/2$, let P_4 and P_5 share B using 1 cut while P_1, P_2, P_3 share A using 3 cuts.

(iii) If $\mu_3(B) > 1/2$, then $\mu_2(B) \leq 1/2$, since at least $(5-1)/2 = 2$ non-cutters think $\mu(B) \leq 1/2$. Let P_3, P_4, P_5 share B using 3 cuts, while P_1, P_2 share A using 1 cut.

9.2. Proceeding as in Exercise 9.1:

(i) If $\mu_5(B) \leq 2/8$, let $P_1, \cdots, P_5$ share A using 5 cuts, while P_6 takes B.

(ii) If $\mu_5(B) > 2/8$ and $\mu_4(B) \leq 1/2$, let P_5 and P_6 share B using 1 cut, while P_1, P_2, P_3, P_4 share A using 4 cuts.

(iii) If $\mu_4(B) > 1/2$, then $\mu_3(B) \leq 1/2$. Let P_4, P_5, P_6 share B using 3 cuts, while P_1, P_2, P_3 share A with 2 cuts using the algorithm used to show $M(3,2) \geq 1/4$. In the last group each gets at least $1/4 \times 1/2 = 1/8$.

9.3. Since $D(3) = 3$, the values through $n = 3$ are convex. Proceeding by induction and asssuming values for $1, 2, \cdots, n-1$ are convex we get

$$\begin{aligned} D(n) &= (n-1) + D(\lfloor n/2 \rfloor) + D(\lceil n/2 \rceil) \\ D(n-1) &= (n-2) + D(\lfloor (n-1)/2 \rfloor) + D(\lceil (n-1)/2 \rceil). \\ \text{So } D(n) - D(n-1) &= 1 + D(\lceil n/2 \rceil) - [D(\lfloor (n-1)/2 \rfloor) \\ &\quad - D(\lceil (n-1)/2 \rceil) + D(\lfloor n/2 \rfloor)] \\ &= 1 + D(\lceil n/2 \rceil) - D(\lceil n/2 \rceil - 1). \end{aligned}$$

(check by cases for n even or odd)

But $D(\lceil n/2 \rceil) - D(\lceil n/2 \rceil - 1) \geq D(\lceil (n-1)/2 \rceil) - D(\lceil (n-1)/2 \rceil - 1)$ because of the convexity of values for $1, 2, \cdots, n-1$. So $D(n) - D(n-1) \geq D(n-1) - D(n-2)$, which implies convexity is preserved by the value $D(n)$.

9.4. If n is even and $t < n/2$, then with $\lfloor n/2 \rfloor = \lceil n/2 \rceil = n/2$ we have $G(n/2) \leq 1/2\ G(t) + 1/2\ G(n-t)$ since $n/2 = 1/2\ (t + (n-t))$. So $G(\lfloor n/2 \rfloor) + G(\lceil n/2 \rceil) = 2G(n/2) \leq G(t) + G(n-t)$ as required.

If n is odd, then $n/2$ is the midpoint of both $[\lfloor n/2 \rfloor, \lceil n/2 \rceil]$ and $[t, n-t]$.

If $t < n/2$, since values of G are convex, the segment joining $(t, G(t))$ and $((n-t), G(n-t))$ lies above the segment joining $(\lfloor n/2 \rfloor, G(\lfloor n/2 \rfloor))$ and $(\lceil n/2 \rceil, G(\lceil n/2 \rceil))$.

Thus $1/2\ (G(\lfloor n/2 \rfloor) + G(\lceil n/2 \rceil)) \leq 1/2\ (G(t) + G(n-t))$. So again, the best bound of $G(n)$ is obtained by choosing $t = \lfloor n/2 \rfloor$.

Chapter 10

10.1. (a) $A - (A \cap B)$ and $B - (A \cap B)$ would still satisfy the given conditions if A and B overlap.

(b) **Step 1.** Have P_1 and P_2 divide $X - (A \cup B) = C \cup D$ in ϵ near-exact portions in the ratio $\alpha_1 : \alpha_2$, where ϵ is chosen small with respect to a and b.

Step 2. Have P_1 divide $A = A_1 \cup A_2$ so that $\mu_1(C \cup A_1) = \alpha_1$ and $\mu_1(D \cup A_1) = \alpha_2$.

Step 3. Have P_2 divide $B = B_1 \cup B_2$ so that $\mu_2(C \cup B_1) = \alpha_1$ and $\mu_2(D \cup B_2) = \alpha_2$. The portions $C \cup A_1 \cup B_1$ and $D \cup A_2 \cup B_2$ are exact in the ratio $\alpha_1 : \alpha_2$.

10.2. Suppose there are portions A_i, $i = 1, 2, \cdots, n$, with $\mu_i(A_i) = a_i > 0$ while $\mu_j(A_i) = 0$ for $j \neq i$. (Of course this is a very strong condition.) As before, if they are not already disjoint, replace the pieces with $A_i - (\cup_{j \neq i} A_j)$. The division process is similar to that of Exercise 1.

Chapter 11

11.1. Any simple fair division method for two players is automatically envy-free.

11.2. Note that at most $d+h-1$ pieces are trimmed in Steps 1–4. These pieces yield the excess pieces that are cut in Step 7, and each excess may be considered a single piece by the One Cut Suffices Principle. Hence, the following are upper bounds on the cuts at each step.

Step 1: $t + d + h - 1$

Step 2–4: $(2d + h - 2)(h + 1)/2$

Step 6: $d + h - 1$

Step 7: $(t + d + h - 1)(d + h - 1)$

Since we may suppose $t \geq d \geq h$, the bound follows by adding these numbers.

11.3. Suppose $n + 1 > 2^{r-1}$. We want to show that any finite algorithm will require at least r cuts. Since the share S given to P_1 must satisfy $\mu_1(S) \geq 1/(n+1)$ and $\mu_2(S) \leq 1/(n+1)$, there must be a single piece D in S with $\mu_1(D) \geq \mu_2(D)$. If at any stage a piece B is cut with $\mu_1(B) < \mu_2(B)$, then D cannot be a subset of B since μ_1 and μ_2 may be proportional on B. Hence piece D will result from a sequence of cuts of the form:

$X = A_0$

A_0 is cut into $A_1 \cup B_1$

A_1 is cut into $A_2 \cup B_2$

...

A_{j-1} is cut into $A_j \cup B_j$ with $D = A_j$.

(Other cuts on the B_i may occur, but we will count only the cuts that produce piece D.) From above we must have $\mu_1(A_i) \geq \mu_2(A_i)$ for $1 \leq i \leq j$.

We will show that there is a measure, μ_2, with $\mu_2(A_j) \geq 1/2^j$ giving $1/(n+1) \geq \mu_2(D) = \mu_2(A_j) \geq 1/2^j$. Thus $2^{r-1} < n + 1 \leq 2^j$ so that $j > r - 1$, and at least r cuts are required for this measure.

If the initial steps of the algorithm instruct either player to cut halves, there may be agreement on the pieces, in which case $\mu_2(A_i) = 1/2^i$, until someone is instructed to cut other than halves. So let us assume it is at the kth cut that the algorithm first instructs one of the players to cut non-halves from A_{k-1}. Then we will have $A_{k-1} = L \cup S$ where the cutter thinks L is strictly larger than S. Regardless who the cutter is, it may be the case that the evaluations have the form $\mu_1(L) = \alpha_k, \quad \mu_2(L) = \alpha_k - \varepsilon, \quad \varepsilon > 0$ (to be chosen later), and $\alpha_k - 1/2^k = \eta > 0$. It follows that $\mu_1(S) < \mu_2(S)$, so D cannot be a subset of S and $L = A_k$.

Proceeding by induction, suppose we are at a stage after n cuts, $n \geq k$, where $\mu_1(A_n) = \alpha_n$ and $\mu_2(A_n) = \alpha_n - (n-k+1)\varepsilon$, (which is the case for $n = k$). Now either P_1 or P_2 will be instructed to cut A_n into two pieces L_{n+1} and S_{n+1} where the cutter sees L_{n+1} at least as large as S_{n+1}. The evaluations could take the form given in the following table:

	L_{n+1}	S_{n+1}
P_1	α_{n+1}	$\alpha_n - \alpha_{n+1}$
P_2	$\alpha_{n+1} - (n-k+2)\varepsilon$	$\alpha_n - \alpha_{n+1} + \varepsilon$

Necessarily D comes from $L_{n+1} = A_{n+1}$ so that the induction statement holds for $n+1$.

For $n \geq k$, if P_1 cuts A_n, then $\alpha_{n+1} \geq \alpha_n - \alpha_{n+1}$ so that $\alpha_{n+1} \geq \alpha_n/2$. If P_2 cuts A_n, then $\alpha_{n+1} - (n-k+2)\varepsilon \geq \alpha_n - \alpha_{n+1} + \varepsilon$ so that $\alpha_{n+1} \geq 1/2(\alpha_n + (n-k+3)\varepsilon) > \alpha_n/2$. Hence when $n = j$, $\mu_1(A_j) = \alpha_j \geq (1/2)\alpha_{j-1} \geq \cdots \geq (1/2^{j-k})\alpha_k = (1/2^{j-k})(1/2^k + \eta) > 1/2^j(1+\eta)$. Finally, $\mu_2(A_j) = \alpha_j - (j-k+1)\varepsilon \geq \alpha_j - j\varepsilon \geq 1/2^j(1+\eta) - j\varepsilon \geq 1/2^j$ if $\varepsilon \leq \eta/(j2^j)$.

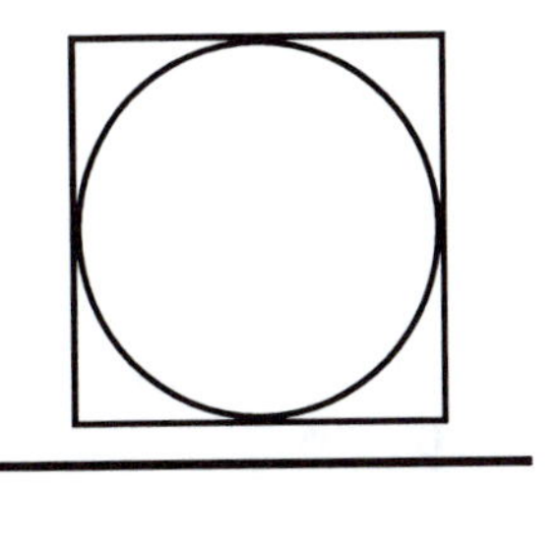

References

[Ale] P.S. Aleksandrov. *Combinatorial Topology*. Vol. 1, Graylock Press, Rochester, N.Y., (1956).

[Alo] N. Alon. Splitting necklaces. *Advances Math.*, **63** (1987), 247–253.

[Aus] A.K. Austin. Sharing a cake. *Math. Gazette*, **66**, 437 (1982), 212–215.

[AS] A.K. Austin and W. Stromquist. Commentary. *Am. Math. Monthly*, **90** (1983), 474.

[BY] M.L. Balinski and H.P. Young. *Fair Representation: Meeting the Ideal of One Man One Vote*. Yale University Press, New Haven (1982).

[Bar1] J. Barbanel. Super envy-free cake division and independence of measures. *J. Math. Anal. Appl.*, **197** (1996), 54–60.

[Bar2] J. Barbanel. On the possibilities for partitioning a cake. *Proc. Am. Math. Soc.*, **124**, 11 (1996), 3443–3451.

[Bar3] J. Barbanel. Game-theoretic algorithms for fair and strongly fair cake division with entitlements. *Colloquium Math.*, **69** (1995), 59–73.

[BT] J. Barbanel and A. Taylor. Preference relations and measures in the context of fair division, *Proc. Am. Math. Soc.*, **123**, 7 (1995), 2061–2070.

[Bec] A. Beck. Constructing a fair border. *Am. Math. Monthly*, **94** (1987), 157–162.

[Ben] S. Bennett, et al. *The Apportionment Problem*. COMAP. Arlington, MA (1986).

[Ber] V. Bergstrom. Zwei Sätze über ebene Vectorpolygone. *Hamburgische Abhandlungen* **8** (1930), 205–219.

[Bla] J.H. Blau. The existence of social welfare functions. *Econometrica*, **25** (1957), 302–313.

[BT1] S.J. Brams and A.D. Taylor. A note on envy-free cake division. *J. Comb. Theory*, A, **70** (1995), 170–173.

[BT2] S.J. Brams and A.D. Taylor. An envy-free cake division protocol. *Am. Math. Monthly*, **102** (1995), 9–18.

[BT3] S.J. Brams and A.D. Taylor. *Fair Division: From Cake Cutting to Dispute Resolution*, Cambridge University Press, Cambridge, UK (1996).

[BTZ1] S.J. Brams, A.D. Taylor, and W.S. Zwicker. Old and new moving-knife schemes. *Math. Intelligencer*, **17**, 4 (1995), 30–35.

[BTZ2] S.J. Brams, A.D. Taylor, and W.S. Zwicker. A moving-knife solution to the four-person envy-free cake division problem. *Proc. Am. Math. Soc.*, **125** (1997), 547–554.

[CL] G. Chartrand and L. Lesniak. *Graphs and Digraphs*, 2nd Ed. Wadsworth and Brooks/Cole, Monterey, CA (1986).

[COM] COMAP. *For All Practical Purposes*, 3rd Ed. W.H. Freeman, New York (1994).

[Dub] L.E. Dubins. Group decision devices. *Am. Math. Monthly*, **84** (1977), 350–356.

[DS] L.E. Dubins and E.H. Spanier. How to cut a cake fairly. *Am. Math. Monthly*, **68** (1961), 1–17.

[EHK] J. Elton, T. Hill, and R. Kertz. Optimal-partitioning inequalities for nonatomic probability measures. *Trans. Am. Math. Soc.*, **296** (1986), 703–725.

[EM] J.R. Evans and E. Minieka. *Optimization Algorithms for Networks and Graphs*, 2nd Ed. Marcel Dekker, New York (1992).

[EP] S. Even and A. Paz. A note on cake cutting. *Discrete Appl. Math.*, **7** (1984), 285–296.

[Fin] A.M. Fink. A note on the fair division problem. *Math. Mag.*, (1964), 341–342.

[Gal] D. Gale. Mathematical entertainments. *Math. Intelligencer*, **15**, 1 (1993), 48–52.

[Gar] M. Gardner. *aha! Insight, Scientific American Inc.* W.H. Freeman, New York (1978), 123–124.

[GW] C.H. Goldberg and D.B. West. Bisection of circle colorings. *SIAM J. Algebraic Discrete Methods*, **6** (1985), 93–106.

[Guy] R.K. Guy. The strong law of small numbers. *Am. Math. Monthly*, **95** (1988), 697–712.

[Hil1] T. Hill. Determining a fair border. *Am. Math. Monthly*, **90** (1983), 438–442.

[Hil2] T. Hill. Partitioning general probability measures. *Ann. Prob.*, **15** (1987), 804–813.

[Hil3] T. Hill. A sharp partitioning-inequality for non-atomic probability measures based on the mass of the infimum of the measures. *Prob. Theory and Related Fields*, **75** (1987), 143–147.

[Hil4] T. Hill. Stochastic inequalities. *IMS Lecture Notes*, **22** (1993), 116–132.

[Hiv] W. Hivel. Dividing the spoils. *Discover*, **16** (1995), 49–57.

[KK] V.M. Kadets and M.I. Kadets. *Rearrangements of Series in Banach Spaces, Trans. of Math. Monographs*, (86). American Mathematical Society, Providence, RI (1991).

[Kna] B. Knaster. Sur le problème du partage pragmatique de H. Steinhaus. *Ann. Soc. Polonaise Math.*, **19** (1946), 228–231.

[Kuh] H.W. Kuhn. *On Games of Fair Division, Essays in Mathematical Economics in Honor of Oskar Morgenstern* (Martin Shubik, ed.). Princeton University Press, Princeton, NJ (1967).

[LW] J. Legut and M. Wilczýnski. Optimal partitioning of a measurable space. *Proc. Am. Math. Soc.*, **104** (1988), 262–264.

[Lew] A.A. Lewis. Aspects of fair division. Rand Corporation, **P-6475** (1980).

[LoWe] C. Long and W. Webb. Analysis of the Euclidean and related algorithms, in *Applications of Fibonacci Numbers* (7). Kluwer Academic Publishers, Dordrecht, NL (to appear).

[Luc] W.F. Lucas. Elementary fair division schemes. *Math. Notes*, Washington State University, **21**, 1 (1978), 1–4.

[MRW] K. McAvaney, J. Robertson, and W. Webb. Ramsey partitions of integers and fair division. *Combinatorica*, **12** (1992), 193–201.

[Oli] D. Olivastro. Preferred shares. *The Sciences*, (1992), March/April, 52–54.

[Osk] R. Oskui. Dirty work problem. Preprint.

[Pou] W. Poundstone. *Prisoner's Dilemma: John von Neumann, Game Theory, and the Puzzle of the Bomb*. Doubleday, New York (1993).

[Reb] K. Rebman. How to get (at least) a fair share of the cake, in *Mathematical Plums* (Ross Honsberger, ed.) *The Mathematical Association of America*, (1979), 22–37.

[RW1] J. Robertson and W. Webb. Minimal number of cuts for fair division. *Ars. Comb.*, **31** (1991), 191–197.

[RW2] J. Robertson and W. Webb. Approximating fair division with a limited number of cuts. *J. Comb. Theory*, A, **72**, 2 (1995), 340–344.

[RW3] J. Robertson and W. Webb. Extensions of cut and choose fair division. *Elem. Math.*, **52** (1997), 23–30.

[RW4] J. Robertson and W. Webb. Near exact and envy-free cake division. *Ars. Comb.*, **45** (1997), 97–108.

[RW5] J. Robertson and W. Webb. Asymptotic values for the number of Ramsey partitions of integers. Preprint.

[Saa] T.L. Saaty. *Optimization in Integers and Related Extremal Problems*. McGraw-Hill, New York (1970).

[Sin] E. Singer. Extensions of the classical rule "Divide and Choose." *Southern Economic J.*, **28** (1962), 391–394.

[St] I. Stewart. Fair shares for all. *New Scientist*, **146** (1995), 42–46.

[Ste1] H. Steinhaus. The problem of fair division. *Econometrica*, **16** (1948), 101–104.

[Ste2] H. Steinhaus. Sur la division pragmatique. *Econometrica* (supplement), **17** (1949), 315–319.

[Ste3] H. Steinhaus. *Mathematical Snapshots*, 3rd. Ed. Oxford University Press, New York (1969), 70–72.

[Str] W. Stromquist. How to cut a cake fairly. *Am. Math. Monthly*, **87** (1980), 640–644.

[Su] F. Su. Cake-cutting, chore-division, and rental harmony. Preprint.

[SW] W. Stromquist and D.R. Woodall. Sets on which several measures agree. *J. Math. Anal. Appl.*, **108** (1985), 241–248.

[Tab] S.L. Tabachnikov. Considerations of continuity. *Quantum*, **1**,2 (1990), May, 8–12.

[Tay] A. Taylor. Algorithmic approximations of simultaneous equal division. Preprint.

[Tes] L.S. Tesfatsion. Fair division with uncertain needs and tastes. *Social Choice and Welfare*, **2** (1985), 295–309.

[Tuc] A.C. Tucker. *Applied Combinatorics*, 3rd. Ed. Wiley, New York (1995).

[Urb] K. Urbanik. Quelques theoremes sur les measures. *Fund. Math.*, **41** (1954), 150–162.

[Web1] W. Webb. A combinatorial algorithm to establish a fair border. *Europ. J. Comb.*, **11** (1990), 301–304.

[Web2] W. Webb. How to cut a cake fairly using a minimum number of cuts. *Discrete Appl. Math.*, **74** (1997), 183–190.

[Web3] W. Webb. Envy-free cake division into unequal parts. Preprint.

[Web4] W. Webb. An algorithm for super envy-free cake division. Preprint.

[Woo1] D.R. Woodall. Dividing a cake fairly. *J. Math. Anal. Appl.*, **78** (1980), 233–247.

[Woo2] D.R. Woodall. A note on the cake-division problem. *J. Comb. Theory*, A, **42** (1986), 300–301.

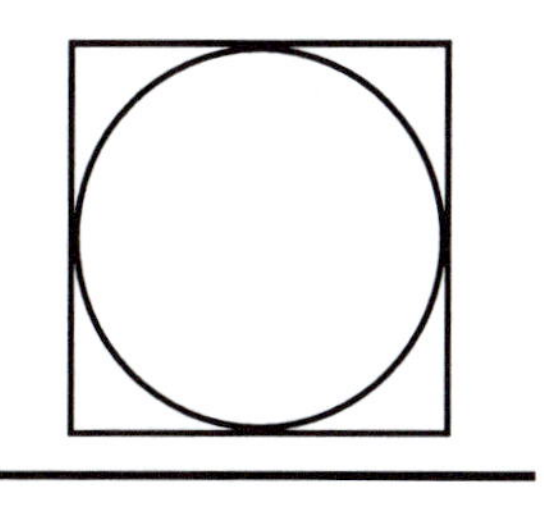

Index